W0246446

The Impact and Value of Science

First published in 1945, *The Impact and Value of Science* is both a plea and a challenge: a plea for more and more science – not to increase the sum total of technical knowledge nor to extend present material amenities, but in the words of the author for the sake of "mental maturity." It is a challenge to try the method of science. Every man is a scientist and every scientist a useful citizen.

Dr. Hill has been both an industrial and academic scientist but here he is concerned with something much wider than textbook conception of science. He gives a clear answer to those who argue that scientific progress is leading to man's destruction by showing that if the scientific method is applied in "non-scientific" fields – in religion, ethics, politics – man will learn how to use the technical inventions of science as stepping stones in social and economic progress. With space tourism and climate crisis marking the two ends of scientific development in current times, this book is of value to everyone but especially to students of climate change, public policy and ethics.

The Impact and Value of Science

D. W. Hill

Routledge
Taylor & Francis Group

First published in 1945
by Hutchinson's Scientific & Technical Publications

This edition first published in 2022 by Routledge
2 Park Square, Milton Park, Abingdon, Oxon, OX14 4RN

and by Routledge
605 Third Avenue, New York, NY 10017

Routledge is an imprint of the Taylor & Francis Group, an informa business

© 1945 Hutchinson's Scientific & Technical Publications

Publisher's Note
The publisher has gone to great lengths to ensure the quality of this reprint but points out that some imperfections in the original copies may be apparent.

Disclaimer
The publisher has made every effort to trace copyright holders and welcomes correspondence from those they have been unable to contact.

ISBN: 978-1-032-20076-7 (hbk)
ISBN: 978-1-003-26224-4 (ebk)
ISBN: 978-1-032-20108-5 (pbk)

Book DOI 10.4324/9781003262244

THE IMPACT AND VALUE OF SCIENCE

by

D. W. HILL, D.Sc.

HUTCHINSON'S
SCIENTIFIC & TECHNICAL PUBLICATIONS
LONDON : NEW YORK : MELBOURNE : SYDNEY

Second Impression February, 1945.

FOREWORD

THERE have been many books in recent years seeking to popularise or explain science, most of them starting from a premise that I do not believe to be generally accepted. There appears to be agreement, among scientists at least, that the world appreciates and desires the help that science can give. This I do not believe to be true. On the contrary, I suspect strongly that the ordinary man regards the modern miracles of scientific achievement with awe but at the same time with profound distrust. He sees in them a continual threat to his peace of mind, the stability of his world and the safety of his person. Not the least of the advantages of the "good old days" was that there was little science and what there was could speedily be checked with the assistance of the stake and a few faggots. In these softer days we are limited to protests against the uses to which scientific discoveries are put or, at most, to demands from Bishops that a period of prohibition should be imposed on scientific research.

The justification of these pages is therefore that they try to show that science resembles the law in that the good man has nothing to fear from it and that the chief contribution that science and scientists have to make to the world's welfare has never yet been tried. It is the thesis of these pages that scientific thought is the mainspring of scientific advance and that it should be also the guide of scientific application. The aim is to try to dispel the fear of the unknown. Nowhere in them will the reader find a definition of science, so, although I have been warned that definitions are traps for the unwary, I give one here that I hope will describe the science (as distinct from the technology of most popular expositions) that I believe can accomplish all the things it is willing to attempt. I take science to mean the power to think logically, dispassionately, impersonally, objectively and thoroughly according to a definite pattern that can be consciously adopted and taught ; to extend knowledge by ordered experiment ; and perhaps one should add, to act fearlessly on the conclusions reached. If this seems to imply that in my view science stands for all that is best in human thought, at all events I have nowhere suggested that these qualities are exclusive to the natural sciences but only that in them they have been systematised so that they can be passed on.

The passages from the Moffatt Bible are quoted by permission of the owners of the copyright, Messrs. Hodder & Stoughton, Ltd.

March, 1944   D. W. H.

A*

CONTENTS

THE SCIENTIFIC OUTLOOK

*"The scientist does not study nature because it is useful ;
he studies it because he delights in it and he delights in it
because it is beautiful."*

HENRI POINCARÉ.

To say that we live in a scientific age is to be trite. We are
satiated with reminders that the sciences are the controlling
factors in modern civilisation ; that they are the basis of industry,
the foundation of wealth, the bulwark of comfort, the safeguard
of health and the terror of war ; that, in fact, without them the
world as we know it would cease to exist. The popular expositions of science to which we have become accustomed usually
remind us, however, only of the sensational achievements of our
civilisation, its motor cars, aeroplanes, wireless communications,
synthetic fibres and chemical industries, forgetting that these
things are products only—the shadow and not the substance ;
and this habit has led to the idea of scientists as technicians—
rather super-technicians perhaps, but technicians nevertheless,
and to science as a collection of technologies. There is no shame
in being a technician, particularly a good one, but it is a misunderstanding of the very essence of science so to limit its scope.
So narrow a view fails to appreciate the fundamental nature of
science.

In stressing its technological value we say, in effect, that it is
important because of what it does, losing sight thereby of the
fact that its real significance rests upon what it is. We are
anxious to employ the sciences to enable us to control and
exploit to our own ends the forces of nature, but the first object of
science is not to control or to exploit but to understand. All
the sciences are branches of one great tree rooted in understanding.
Just as no tree springs fully grown from the earth, so some
branches of this tree have become great boughs supporting lesser
branches, twigs and multitudinous leaves, while others have as
yet shown themselves only as incipient buds. But, branch or
bud, they all draw their life from the same root. The root of
chemistry is the understanding of the materials of the universe ;
of physics, the understanding of the forces around us ; of
psychology, the understanding of man's own nature. It is a

7

very old root. The desire to understand is a universal charac-teristic and in so far as men foster their desire to know they become potential scientists. They may not be good scientists and they may not become professional ones. They may be musicians, painters, theologians, barristers, costing clerks, tram drivers or lumbermen who have been endowed by nature with a child-like all-devouring curiosity. It is here that science comes into its own. In its contribution to man's desire for understanding it stands majestically alone, for it "has brought instruments of precision into the domain of thought which can be used more easily and more safely than any other."[1] The greatest claim of science is not that it has produced synthetic plastics, Dewey classifications, motion studies or almost infallible means of navigation, but that it has developed a method of thought that has surpassed all previous methods. It is the inheritor of the Greek tradition. The ancient Greeks, who forged intellectual weapons for all to use and were indeed the first scientists, were they alive to-day, would embrace enthu-siastically the science that has so extended their thought and philosophy. It has outstripped them in a race in which they were the acknowledged leaders. It has provided an intellectual armoury mightier even than theirs. Yet it is the same kind of armoury. Professor Gilbert Murray has suggested that a first acquaintance with Greek thought might lead to boredom "because it is all so normal and truthful ; so singularly free from exaggera-tion, paradox, violent emphasis."[2] If this is not a description of scientific thought then none exists.

It might well be asked why, nowadays, the division between classicist and scientist is so pronounced, even to the point of antipathy. A gulf seems to be fixed between science and classical studies, the blame for which must rest equally upon the classicists to whom the acceptance of tradition has seemed of greater importance than the pursuit of truth, and the scientists who, engrossed in the utilitarian value of their study, have lost sight of it as a mental discipline.

Science is undoubtedly important for what it has done and it will be impossible in subsequent chapters to avoid references to its technological achievements which both provide illustrations of its power and constitute, because of their repercussions upon both the individual and the community, one reason why we must appreciate the deeper significance of science and learn to

[1] A. Smithells. *From a Modern University.*
[2] *The Legacy of Greece.*

use it to the full extent of its value. It seems inevitable that acquaintance with the sciences must increase if only because of their technological and utilitarian value. In the world we now know we cannot afford to neglect them and as we become more conversant with them, so it may be expected that their underlying discipline will seize hold upon us. There is, however, a real danger that like the Romans, who were never really interested in science except for what it brought them, we may take the material benefits and neglect the spiritual force. A large part of the present world chaos may safely be attributed to the fact that we have learned largely to control nature, to exploit natural laws to our own ends and to let loose natural forces that are almost without restraint, while disregarding the spirit and discipline of the science by which we are informed. This is as true of professional scientists as of laymen and no man has earned the right to be called a scientist, however mighty his intellectual achievements or distinguished his academic honours, who has not submitted his mind and outlook to the discipline of his learning.

Herein lies the chief danger of regarding science as a collection of technologies. The practical arts which may have benefited from the application of the scientific method remain technologies and their exponents technicians ; the chemist, the physicist, the doctor, the engineer are all technicians in so far as they are practical men utilising to the best advantage the tools of their professions. Only in so far as they are informed by the spirit of science are they scientists. In this sense, many professional scientists are no more than technicians. The barrister skilled in the ways of the Courts, the preacher skilled in pulpit arts, the politician skilled in the finesse of public life, the financier skilled in the ways of money, are technicians of a high order but nobody thinks of calling them scientists, and in the same sense there is no intrinsic reason why a man who has acquired an ability in the natural sciences should of necessity be a scientist. One thing alone can make him that. Nevertheless he possesses an advantage in that the scientific method has found the most receptive soil and borne its finest fruit in the study of natural phenomena so that the student of nature has an opportunity of becoming a scientist by virtue of a contact which is denied to the majority of men.

It is evident that science and technology are not synonymous. Technology, in fact, thrived for many centuries before science was organised. Tubal Cain far back in history was an "instruc-

tor of every artificer in brass and iron." During the dark ages
and into modern times the craftsman and artisan persisted.
When science was reborn in the seventeenth century, investigators sought to profit by the skill of craftsmen and to-day
scientists in their secluded laboratories or ensconced in their
observatories are dependent for their physical tools upon the
products of craftsmen often with little or no scientific training.
Skilled glass blowers, instrument makers and mechanics are the
servants of the sciences without whom few of the observations
upon which scientific advance has been based could have been
made. The telescope of the astronomer, the microscope of the
bacteriologist, the precision balance of the chemist, the
micrometer of the engineer, the X-ray tube or amplifying valve of
the physicist are the products of the craftsman's long apprenticeship. Scientists may and usually have contributed to the
understanding of the tools of their professions and thus to
improvements in design, as, for example, in the technique of
microscopy, which starting with the magnifying glass passed
through the compound lens, the optical microscope in various
stages of perfection and now, finally, the electron microscope ;
but it is here that we are impressed by the essential difference
between science and technology. The former may invent and
improve and extend but this is not its first concern. Its pre-
eminent business is to understand and by its understanding to
extend all knowledge. The latter is concerned, not so much to
understand as to produce and use to the best advantage.

It is interesting to recall that not so long ago the distinction
between the different sciences which we now accept did not
exist. The great scientists of the seventeenth, eighteenth and
early nineteenth centuries were not specialists. Their interests and
activities ranged over the whole field of observable natural
phenomena, to all of which they applied the same kind of critical
thought. Sir Christopher Wren, for example, best known for
his architectural genius, first made a name for himself as a
physiologist, from which he graduated to astronomy, becoming
Savilian Professor in this subject at Oxford, before turning to
architecture. As information was collected and knowledge
grew, it was inevitable that sub-division should occur and men
who had been anatomists, chemists, metallurgists and physicists
combined, were forced into making a selection of the information
that appealed most to them and so became specialists. But
through all the sciences and all their branches runs the con-
necting chain of a similar procedure and identical mental habit.

Sciences, in fact, differ only in the technique of observation. Always it is the attempt to apply the scientific method that has built them up and the day seems not far distant when it will be universally recognised once more that though there may be one technology of physics, another of chemistry, another of botany and another of physiology, yet the underlying principles are natural relationships interpreted and used in the same way in all sciences for the understanding of differently described phenomena. "There is one glory of the sun and another glory of the moon, and another glory of the stars ; for one star differeth from another star in glory," says St. Paul. So it is with science. The technology of astronomy is not that of bacteriology, nor is the technology of zoology that of geology. There are many technologies but one science infuses them all, inspirits them all, guides and develops them all, explaining, co-ordinating and understanding them all.

The technology of the artisan is not very different from that of the scientist. The former, whether he be fitter, carpenter or electrician, learns during his apprenticeship to use the tools of his trade and to become a competent workman able to take a job and carry it through to completion. In the same way, the scientist during his formative years and afterwards by long experience becomes a skilled user of the tools of his trade. If he becomes no more he remains a specialised artisan. If a chemist, he may become a competent analyst, familiar with the elementary mathematics necessary for the task, skilled in the use of the balance and the burette, and adept at the tricks of the trade which make for accuracy. If an astronomer, he may become a skilful observer of the skies, if an anatomist, a skilful dissector, if a physicist, a skilful microscopist ; but in all these things he is an artisan as surely as the skilled fitter or tool-maker who works to thousandths of an inch. The additional something that makes him more than an artisan and raises him to the level of a scientist is the ability to initiate, to correlate, to synthesise, to interpret, and the possession of that inward force which is the spirit of science.

In practice, however, the term technology is often restricted to the industrial arts. The scientist applying his knowledge to the establishment of new or improved products or processes in industry is referred to as a technologist. And so he is but he is not therefore less of a scientist than his academically minded brother who abhors industry and all useful arts. The successful scientist must also be a competent technologist wherever he may

be situated and whatever application he may make of his knowledge and skill.   If he fails as a technologist he fails in an important professional quality.

But technology is only one aspect of the application of science,
and the spirit of science goes deeper than its technical applications.   This spirit infused into it by generations of men with a
passion for truth has resulted in methods of reasoning that permit
us to judge all questions.   In its finest expression, self-interest,
personal bias and prejudice are alien to it.   It is impersonal and
objective in its outlook, uninfluenced by the world's hatred
and unseduced by its prizes.   Under no circumstances can science
see black as white or the worse as the better.   Tyranny, brutality
or expediency may distort its uses but they cannot finally alter
its course, for it has given its true disciples an ideal and hardened
them by its severe discipline to follow after it.   The scientist
cannot deny himself ; into whatever field of human endeavour
he may wander, his mental tools will not lead him astray.   Provided he maintains the unswerving integrity, the steadfast self-
control and the urge to strip off the specious trappings in order
to penetrate to the fundamentals of any question with which he
may be confronted, he will not be permitted to go far wrong in
his judgments.

The precincts of reason may to some seem harsh and forbidding, but there is something more.   The great scientists have
never been afraid to indulge their imaginations.   They may have
been controlled by a vast array of established facts which they
could not dispute, limited perhaps by the self-imposed discipline
of reason, but never quenched and often enough only induced to
burn the more brightly like a flame in a current of invigorating
oxygen.   The joy of creation is the same in science as in art,
literature or music.   It is the medium only that is different and
the true scientist is an artist to the finger-tips, sensitive, responsive, perhaps even temperamental.   In his heart the scientist is
a poet, with eyes to see where other men are blind, ears to catch
the melodies of the universe to which other men are deaf, and a
finger on the pulse of the world when other men are insensitive.

Science, then, is more than technology and is greater than
the sciences.   It is a system of thought, a philosophy and a guide
to maturity.   It is a living thing of joy and beauty intimately
interwoven with the affairs of life and yet distinct from them.
It is a medium of expression in which the imagination has full
play.   In short, the method of science has revealed for the first
time not only the beauties and harmonies of the universe

but also the ingenuity of the human mind which can build such a superstructure as the natural sciences on the foundation of observable facts. It ranges from the electron to the planet and includes in its orbit the animate and inanimate worlds from the bathosphere to the stratosphere. It is not the only thing in the world. It may not be the greatest, but in the realm of thought it stands alone.

SCIENCE AND INDUSTRY

"Science has its roots and has gained its greatest impulse in the avocations of men."
ARTHUR SMITHELLS.

INDUSTRY and the sciences have had a long and intimate connection with each other, although industry and commerce were flourishing long before science had registered its present technical achievements. Out of such industry and commerce the first interest in the natural sciences developed. The earliest scientific investigations were rarely undertaken for the sake of truth alone. They grew out of efforts to solve the problems of daily life. Advances in mathematics enabled the Egyptians and the Babylonians to develop the leading operations in arithmetic and elementary geometry. Astronomical research flourished, possibly from a superstitious interest in astrology, but also for the provision of a calendar for farming, while the Phœnicians used the knowledge so obtained for navigational purposes and may have been the first sailors to steer at night by the aid of the North Star. The skill of the Ancients as mechanics and engineers is testified to by their buildings, and the transportation which they often involved of colossal weights of stone shows an acquaintance with the principles both of the lever and of the inclined plane. Zoology, botany and mineralogy were also studied and a considerable knowledge of medicine was established. By their investigations the Babylonians and the Egyptians accumulated and passed on an enormous store of information which was directed throughout by an intense practical interest.

It was not until the rise of Greek philosophy that there came a class of men who loved knowledge for its own sake. The Greek philosophers were the forerunners of the modern scientists in that they were not content to accept traditional explanations, but their predilection for philosophical speculation without regard for observable facts prevented their becoming a great scientific nation. The divorcement of philosophy from the practical issues of the needs of men introduced an hiatus in the development of civilisation that lasted until Francis Bacon in the seventeenth century reintroduced the spirit of observation and experiment. During this period men were, with few exceptions, true barbarians accepting the material advantages of technologists and artisans without wishing to understand the processes they employed.

Out of the teaching of Francis Bacon as practised by his disciples arose The Royal Society. The Philosophical College, or Invisible College as Boyle called it, had been in existence for some fifteen years when Charles II in 1660 gave his approval to its foundation, and in 1662 the College received the seal to the Charter which formally incorporated "The Royal Society for the Improvement of Natural Knowledge." The Society was a public expression of the new spirit of investigation as set forth in one of its own early minutes which clearly stated that it would hold no opinion "till by mature debate and clear arguments, chiefly such as are deduced from legitimate experiments" its truth had been demonstrated. From its inception the Society was intensely interested in the practical arts and in applied science and committees were organised for the purpose of collecting and arranging the available knowledge in industrial science.

It is not, of course, to be supposed that the Royal Society began immediately to exert a profound influence on industry, nor that the new science began to cast a spell over the practical arts. A new spirit was abroad, it is true, but its influence was limited to a very narrow circle and a considerable time elapsed before the products of the new thought began to show themselves in industry. In the seventeenth and eighteenth centuries industry was carried on by merchants who were also employers of labour, occasionally to the extent of as many as two or three thousand people. The employees worked in their own homes, were largely their own masters and had few or no formal regulations to obey. The so-called Domestic System continued to operate for the satisfaction of the needs of local communities, chiefly because of transport difficulties. There were no railways, few canals, and roads were often little better than rutted lanes. In consequence, many of the industrial methods in use hardly differed from those employed by the Egyptians, Greeks or Romans for the same purpose .

The inception of the new scientific spirit typified by the Royal

Society and the growing urge for increased production to meet expanding markets were, however, preparing the ground for the seeds of the revolution in industry that was soon to come. The forerunners of great changes may be seen in the spasmodic introduction of mechanical inventions designed to redress the manufacturing balance and usually the outcome of the genius or patient experimentation of one man. The cotton industry provides an excellent example of the trend of the time. In 1738 came the invention by Kay of his mechanical method of propelling the shuttle across the loom. By this means cloth production was greatly speeded up and a demand for more yarn arose, but it was not until 1767 that this demand was met by Hargreaves' "Spinning Jenny," followed in 1779 by Crompton's Spinning Mule. These, in turn, were followed by the introduction of the power loom by Cartwright in 1785.

A period of mechanical and scientific invention in the cotton industry was now well under way and the work of men renowned in scientific as well as industrial circles began to influence its progress. Mechanical invention which had greatly increased the output of yarn and cloth had left the finishing section practically untouched. At the end of the eighteenth and beginning of the nineteenth centuries, Dr. Home of Edinburgh introduced dilute sulphuric acid for neutralising the alkali left in the cloth after scouring, thus doing in less than twenty-four hours what had previously occupied some weeks. This was assisted by the cheapening of the production of this acid by Dr. Roebuck's invention of the lead chamber process.

At the same time, another step forward was being taken by the establishment of synthetic alkalies. Previous to this time, alkali had been obtained either from vegetable ashes or from a few natural deposits (e.g. Trona in Egypt), but here also demand was beginning to exceed supply. In France the position was particularly acute, since the importation of alkali could be, and frequently was, prevented by enemy blockade. In 1775, therefore, the French Royal Academy of Science offered a prize for the best practical method of converting salt to alkali. National self-sufficiency is not a new feature of world economics. The prize was won in 1790 by Nicholas LeBlanc, who started a factory for the purpose at St. Denis, near Paris. It was not long before the LeBlanc alkali process was being worked in England, the first works being started at Walker-on-Tyne in 1796. The deposits of salt in Cheshire, Lancashire, Derbyshire and Durham provide an almost inexhaustible supply of salt and so formed the basis of the British alkali trade. It is not surprising, there-fore, that James Muspratt should establish alkali works near Liverpool, in conformity with the principle of the location of industry according to the distribution of raw materials which was to become so prominent a feature of nineteenth century industrial development and which established the north of England, with its adjacent coal, iron, salt and limestone, as the seat of industry. The LeBlanc process and the establishment of alkali works along the Mersey basin also revolutionised the soap industry. Soap had been an article of commerce with the Phœnicians as early as 600 B.C., but the LeBlanc process, coupled with the researches of Chevreul on the structure of the fats, created a modern industry with world-wide ramifications and a capital running into many millions.

By the beginning of the nineteenth century the cotton indus-try had been completely revolutionised in the space of about seventy years. All the changes were technical and dependent upon either the inventive genius or the scientific skill of indi-viduals, but the rapidity with which they were adopted and became an integral part of the industrial processes is a testimony to the potent economic forces dictating the changes and to the flexible and courageous outlook of the industrial leaders of the time. These advances placed the British cotton industry in the forefront of competitors for world markets and gave to it a favoured position that enabled it to take advantage of every commercial opportunity, and by 1835 cotton goods formed almost fifty per cent of the country's total exports.

This high proportion later declined, not because of any falling off in the cotton industry, which, indeed, continued to expand, but because of the advance of the metallurgical indus-tries, for the happenings in the cotton industry were sympto-matic of industrial changes throughout the country. In the iron and steel industry in particular, Great Britain became supreme for a time, mainly as a result of the introduction of technical improvements. This industry forms a strong contrast to the cotton industry, for whereas the introduction of new ideas into the former has been continuous right up to and into the present century, the latter in the early eighteen hundreds possessed essentially the same form of organisation and much the same machinery in nearly all essentials as it has to-day.

Before the eighteenth century, charcoal-fired furnaces were the source of iron. The increasing demand for the metal stimu-lated the search for some other fuel, and the discovery by the

Darbys that coke could be used successfully to produce iron for castings produced a rapid expansion of the foundry trade. In 1784 Henry Cort perfected the puddling furnace, and with the introduction in 1790 of Homfray's method for the removal of the silicon from the pig iron the malleable iron industry was created which was responsible for Great Britain's supremacy up to 1875. The next great technical advance was again conditioned by economic need. There was a small Scottish iron industry, but the Lanark coal could be used only with difficulty to reduce the local iron stone. Between the years 1828 and 1831, however, suitable conditions were found for using the coal and upon this discovery arose a new and greater Scottish iron industry. Science and technology are, however, no respecters of persons and this discovery also made possible the rise of the American iron industry. Wrought iron remained the principal product of the industry until the invention in 1856 of the Bessemer process for steel. Shortly after this came the Siemens-Martin open hearth process, and with the advantages that steel possessed over wrought iron for many purposes it began gradually to supersede it in many fields. The change over was not rapid, for the simple reason that both the Bessemer Converter process and the Siemens-Martin process required, for their successful working, ores of low phosphorus content and these were not common. The hematite ores of Lancashire and Cumberland were non-phosphoric, and in consequence a steel industry grew up there, but, in addition, the superiority of Great Britain as a shipping nation enabled her to import the non-phosphoric ores of Sweden and Spain. This advantageous position was not, however, to be long maintained, for in 1879 Gilchrist and Thomas devised a basic lining for Bessemer converters that would take up the phosphorus from the iron and so permit good quality steel to be made from phosphoric ores. Once again competitors were not slow to take advantage of discoveries made in Great Britain, and both America and Germany began to exploit their ores rich in phosphorus by the basic slag method and before the end of the century they had overtaken this country as steel producers. Invention and discovery continued apace into the present century and gave us the alloy steels of manganese, silicon, chromium, nickel and tungsten, each possessing qualities peculiar to itself fitting it for particular purposes.

These two illustrations have been given in some detail, to show that the staple trades of Great Britain—coal mining, iron and steel, textiles, engineering—were built up and prospered on a basis of scientific and technical discovery which commenced at about the same time as men were learning the new scientific thought and coincided with the rise of the physical sciences. It would be foolish to try to estimate to what extent they were the outcome of the new thought, but it is evident that scientists played a not inconsiderable part in the building and that both science and industry benefited from the advances in each and the liaison between them.

Not only did these ancient industries benefit, but many industries that we now consider old are really of comparatively recent origin and owe their existence to scientific discovery. The gas industry, for example, is barely one hundred and fifty years old. Inflammable gas from coal was known before that time, but in 1792 William Murdoch conducted it through an iron pipe to his house in Redruth. The gas industry itself may be said to have been founded in 1810 with the passage of an Act for the formation of a statutory company for supplying coal gas to London. The company, which was incorporated a further two years later as the Gas Light and Coke Company, must have flourished from the beginning, for according to Frederick Accum, the chemist to the company, there were already in 1819 about 51,000 gas lights and 288 miles of mains in the metropolis. The Bunsen burner and the incandescent gas mantle, which did so much to render the use of coal gas an efficient proposition and which still form the basis of the majority of its uses, were discovered and perfected in the chemical laboratories of Bunsen and Auer V. Welsbach.

The gas industry forms a link between the ancient staple industries whose courses have been materially modified by science and the new industries that owe their very existence to it. The vast electrical industry, the plastics and synthetic fibre industries, the automobile and aeroplane industries, the dye-stuffs, fine chemicals and drug industries, to mention only a few, were initiated and built up by scientific endeavour, and every industry and every branch of industry are now affected in greater or less degree by the advances in apparently remote scientific fields. Agriculture, architecture and transport are all dependent on science to fulfil their modern destinies.

The research work which has made technical advance possible has been undertaken often enough without reference to immediate problems. An illustration of this may be selected in the twin fields of agriculture and transport. Among the lines of study approved by the Food Investigation Board when it was set up in 1917 was one on the physiology of living and dead foodstuffs, and in the course of this a condition of apples was discovered in which the fruit, outwardly sound, was internally decayed. This condition, known as "Brown Heart," was found to result, not from infection as might at first be supposed, but from partial suffocation. Shortly after, in 1922, the attention of the Board was called to three cargoes of Australian apples that had suffered loss by internal decay. This had already been attributed, without any real basis for the belief, to insect injury in the orchards, but the Board's investigators had no difficulty in recognising "Brown Heart." Examination of the atmosphere in the ship's holds in which the apples had travelled soon showed that the fruit had actually been suffocated by the carbon dioxide they had themselves produced during the voyage.[1]

It is impossible to say what scientific work will be of value in the near or distant future. Who could have predicted that the electrical industry would arise from Faraday's elementary experiments in induction, or the great network of wireless communications that covers the whole world from the early experiments of Helmholtz on electrical waves. All scientific research pays dividends. It is sometimes difficult to convince industrialists of this elementary fact, but the brief and necessarily sketchy outline that has been given here shows to some extent how far industry and commerce are now dependent upon scientific discovery. The financier may claim that industry depends on money, or better still, credit ; the entrepreneur, that it depends on the courage and business ability of individuals ; the trade unionist, that it depends on labour. And in so doing they will all be right, but they will be speaking only a part of the truth, for the scientist may claim that industry is built upon scientific discovery, on the facts of soundly established technical processes. This is the foundation. Money, ability, labour, are all necessary, just as furnishings are necessary to a house before it becomes a home, but the bedrock of modern business is science. It is the only trustworthy secret of success, and it must be recognised that just as science is responsible for the growth of modern industry so with increasing severity of competition only constant and unremitting scientific research will permit any industry to continue to exist, let alone flourish. Sir Henry Tizard, in his Mather Lecture before the Textile Institute in 1929, pointed out that "No industry and no firm within an industry which has

[1] Seventh Annual Report of the Advisory Council for Scientific and Industrial Research, p. 31. Eighth Annual Report, p. 60.

maintained an adequate research department for the continuous improvement and development of its products has ever suffered from long periods of depression. Even in the United States of America one can point to the temporary failure of large and prosperous firms which have neglected to carry through a consistent policy of development and research, but no one can quote a case of failure on the part of any firm which has resolutely pursued such a policy."

The necessity for scientific research cannot be said to have been recognised to any significant extent by industrialists until the present century, although it is true that a few industries realised earlier that the field was so extensive and the possibilities so enormous that only systematic research could hope to cover it adequately without, on the one hand, leaving wide gaps, or, on the other, wasting valuable time on fruitless quests. The early history of the dyestuffs industry in the United Kingdom abounds in names well known in chemical science, the majority of them, alas, foreign, but even this industry, starting out so promisingly, lost the initiative that it possessed by virtue of its outlook and thereby passed to Germany which had a better respect for scientists and the benefits they could provide. The twentieth century, however, has seen some reawakening to the necessity for science. This was undoubtedly hastened by the serious decrease in the value of British exports, by the advent of the last war, and by Government enterprise in granting financial assistance to a scheme for encouraging research. It has been pointed out that in the staple trades of cotton and steel, discovery and invention, particularly in the latter industry, benefited others besides Great Britain. Before the end of the last century both the United States and Germany had overtaken us as steel producers, the dyestuffs industry was fast becoming a German monopoly and by the beginning of the present century Great Britain had lost her predominant position as an exporting nation. In the last thirty years of the nineteenth century the only exports to increase in volume were coal and such manufactures as necessitated only cheap unskilled labour. The remainder of our industries were in danger of being lost through the scientific and engineering ability of Germany and the United States of America.[1]

The early years of the present century seemed to bode better, for the staple trades improved and exports increased. Much of the improvement was, however, accounted for by the increased

[1] W. J. Ashley. *The Tariff Problem.*

exports of the steel and engineering trades financed by British investments abroad, which allowed the borrowers to purchase machinery and transport equipment for their own industrial development. On the other hand, new industries such as the automobile, rubber and electrical industries were beginning to share in the industrial life of the country, but unfortunately the new industries, although important at home, were far behind similar industries abroad. In 1914 Great Britain still depended to a far greater extent than other nations on the staple industries in which she had specialised for a hundred years and these were often being worked inefficiently with semi-obsolete plant compared with other nations, which, starting later, were able to take advantage of improved manufacturing processes that were barred to British industry by the capital already sunk in existing plant. On the whole, therefore, there appeared to be good grounds for pessimism regarding the future, and, in the straits in which it found itself, industry began in a half-hearted fashion to call in the aid of organised industrial scientific research.

The need for scientific intervention for the benefit of industry was also recognised by the Government during this period. As early as 1887 the Imperial Institute was founded in order to co-ordinate and distribute information with the object of encouraging the trade and industries of the Empire. The Institute was reconstituted in 1902, transferred to the Government and began to undertake scientific research related to industry. In 1900 another important step was taken "in the encouragement of organised scientific support for our trades and industries" by the establishment of the National Physical Laboratory. The Laboratory was opened by the Prince of Wales, later King George V, who suggested in his speech that the founding of such an institution showed "that the nation is beginning to recognise that if her commercial supremacy is to be maintained, greater facilities must be given for furthering the application of science to commerce and manufacture." During the early years of the century, too, the new provincial universities were being constituted. They owed their foundation and growth largely to the call of industry for scientific aid, and many of them had established important technological departments.

In 1907 a further important stage in the recognition of the needs of industry was marked by the establishment under a Royal Charter of the Imperial College of Science and Technology, designed "to give the highest specialised instruction and to provide the fullest equipment for the most advanced training and research in various branches of science especially in its application to industry." The Royal College of Science and the Royal School of Mines were transferred to the Imperial College which was provided with a subsidy of £20,000 per annum, thus making it, in contradistinction to the provincial institutions, a national effort.[1]

Judging from the incomes provided for these ventures it seems that the extent of the need was still unknown. It was not until the advent of the first world war that the industrial world realised just how far British industry was behind its foreign competitors. Certain industries then found themselves immediately cut off from essential supplies. Chemical glassware necessary for control in the steel industry was unobtainable. Dyestuffs to the value of £2,000,000 had previously been imported for the textile industry, and an industry with an annual output worth more than £250,000,000 ran the risk of being extinguished for the lack of this essential component. Pharmaceuticals in astonishing quantities had previously been imported from Germany. The pottery manufacturers were threatened with closure by the loss of the Segar cones used as a guide to the firing of china and earthenware. Worst of all to an island community, there was a real fear of starvation and it was this which led to the formation of the Food Investigation Board.

The Food Investigation Board was set up by the Department of Scientific and Industrial Research, which was born under the duress of war for the purpose of "establishing or developing special institutions or departments of existing institutions for the scientific study of problems affecting particular industries and trades." From the beginning it was recognised that the application of science required by the exigencies of war would be no less necessary in peace and the programme of the new Department was formulated accordingly. Besides assisting individual researches the Department embarked on two large schemes. The first of these was the development of co-operative research associations for industry that were to be supported by the industries concerned, and, to an extent dependent on this support, by Government funds. Amalgamation and association were in the air and it was not long before various trade associations had been formed with the sole object of carrying on research for whole industries. The principle involved in the formation of such associations was a simple one. It was accepted that research

[1] First Annual Report of the Advisory Council for Scientific and Industrial Research, 1916.

had much to offer but it was also realised that its cost on a suitable scale would be prohibitive to any but the largest units. In the research associations lay a means of bringing the combined efforts of many minds to bear on the problems of industry at a trifling cost to the individual members of the associations. By 1923 there were twenty-four such associations in existence. Since then others have been formed, while a few of the early ones have ceased to function. It would be a bold man who would maintain that these associations have not fulfilled a tremendously important function in the modern life of British industry and an almost equally bold one who would try to assess the extent of that function.

The second great scheme developed by the new Department arose from the recognition that much research vitally necessary to the welfare of the country hardly falls within the purview of any single industry. Research into the utilisation of fuel is one such. In some form or other every individual in the country is interested in fuel economy and a central Fuel Research Board was therefore established. Research into food problems also falls into this category and the Food Investigation Board to which reference has already been made was therefore founded. The Building Research Board and the Forest Products Research Board followed in due course. These Boards have research stations attached to and under the control of the Department of Scientific and Industrial Research in which investigations are carried out into problems connected with all branches of their interests.

On the face of things it seems therefore as if, after long years of waiting on the doorstep of industry, science is now safely ensconced within the gates and it is desirable to consider in general terms what it has done and what it is likely to do in the future for industry. Already it has transformed the face of the industrial world. Modern transport, the lifeblood of industry, is dependent on it. Not only have the inventions of science made it possible, but it is the skill of technicians that keeps it going. Railways and road transport, the shipping interests and aviation were all born in the brains of scientists and technologists and are maintained by their skill and application. Communications, the delicate nervous system of industrial and world affairs, have spread their network over the face of the globe from the humble beginnings of scientific laboratories. The telephone, telegraph, submarine cable and wireless station are the everyday tributes to the scientists and engineers who made them possible and who keep them in action. These two fields have themselves created other huge new industries—the vast commerce in petroleum is the outcome of the automobile and aeroplane industries. Before their advent natural oil was an extensive article of commerce, but it has now assumed a role in which it controls the destinies of nations. The aeroplane industry stimulated the demand for light alloys and the aluminium industry was born. Aluminium metal which was exhibited as a curiosity costing about £12 a pound at the Paris Exhibition in 1853 was being produced in this country before the present war at the rate of 300,000 tons a year and at a cost of about one shilling a pound. The development of the aluminium industry necessitated in turn a cheap and continuous supply of electricity, and the British Aluminium Company therefore became pioneers of hydro-electric power, establishing works at Foyers and later at Lochabers.

The story of the interrelations of industries could be continued almost indefinitely. The telephone, telegraph and wireless as well as such ventures as the hydro-electric schemes mentioned above, gave a fillip to the new and rapidly growing electrical industry. The debt of industry to science, therefore, is comparable only with the size of industry. Old industries have been revivified, new industries created and the process is continuing at an ever-increasing pace. The organic chemical industries of to-day were founded on scientific researches into the constituents of coal tar, but it is apparent that the organic chemist has found new raw materials in mineral oil and natural gas which under his transforming hand will form the basis of a new chemical industry in the near future. Scarcely had some understanding of the nature of high polymers been reached when new plastics were on the market, not as substitutes for old materials but as products more suitable to their purpose than the natural ones upon which we had previously been dependent. The latest of these is the Nylon family of synthetic fibres, which bid fair by their qualities to replace natural silk in the world's markets. We must expect that new and great industries will arise from the application of scientific knowledge to commerce and the industrialist must watch and keep abreast of the developments in pure science or he and his industry will inevitably decline. The explanation of the financial columns of the newspapers is to be found increasingly in research laboratories. Dividends and stock quotations are dependent on the scientists behind the scenes. It is reported from America that in 1937 forty per cent of the sales of the Du Pont Company were accounted for by twelve new products all developed since 1928, and that, of the nett income of the Union Carbide and Carbon Company, forty-eight per cent came from products made available since 1919.

There is another important aspect of the impact of science on industry. It has redistributed and evened out world resources of raw materials. During the last war Great Britain was dependent to an alarming extent for her nitrogen supplies on imports from the Chile saltpetre beds with their attendant loss of currency and danger in transport. Munitions for war and fertilisers for agriculture hung upon a narrow thread across the South Atlantic and the Chilean nitrate owners benefited accordingly. But every nation has an inexhaustible supply of nitrogen in its atmosphere and the German chemist Haber has shown how to get it into a utilisable form and thus rendered every country independent of the Chilean deposits. New chemical industries have arisen which have made nations independent of imported raw materials. In the textile industry rayon, discovered in Britain, can, if necessary, make nations independent of the American cotton belt ; lanital, discovered in Italy, may render that country independent of the Australian wool clip ; and Nylon, discovered in America, may make us independent of the Japanese silk-worm. At the beginning of the nineteenth century Great Britain's pre-eminence was largely dependent on seemingly illimitable supplies of easily won coal. As the principal source of power it placed the industrial world at our feet. Since then science and technology have made the coal-fields of other nations of comparable importance to our own, and the principal new sources of industrial power—fuel oil and water power—are outside our borders. The relative importance of coal has therefore declined, and we, who used to be the proud possessors of the sole source of power, now see it facing serious competition and have ourselves been reduced to importing enormous quantities of oil. With the advent of the new chemical industries a fresh value has been given to the orientation of raw materials. It may not now be so important as it was at the beginning of the industrial era that coal and iron should be found in close proximity. It may be more important for us that coal and limestone should be adjacent.

The location of industry in the past was dependent to a large extent on superficial natural advantages modified sometimes by the advent of competition from other sources. Thus Japan abandoned the production of cotton in favour of silk when cheap supplies of the former became available from India and America.

B

Industrial centres in the south of England languished when the coal-fields of the north and midlands began to attract industry with the promise of cheap power. The Lancashire cotton industry affords an excellent example of a location of industry that would be hardly likely to occur to-day. Progress in humidification engineering has made it possible to spin and weave cotton in any climate and the advantage of the damp Lancashire atmosphere is thus largely offset. Moreover, even in Lancashire, although the dampness in the valleys where the mills are mostly located is greater than on the hills, and where the mill buildings become impregnated with moisture and so act as a reservoir for humidification, an east wind or a prolonged frost will dry out both the air and the mill buildings. For efficient spinning and weaving artificial humidification is therefore sometimes resorted to even in Lancashire.[1] The localisation of the film industry in California has commonly been attributed to the clarity of the air and the dependable sunshine which make successful outdoor photography possible on most days of the year. At one time it looked as if this natural advantage would preserve for Hollywood a monopoly of the industry. The introduction of Krieg lights and fast films has, however, enabled other centres to establish an industry under artificial conditions. Improved transport facilities, as well as the scientific control of conditions, have added to the freedom of location of industry and largely done away with the predominating influence of natural advantages.

At the same time that science was making nations independent of many raw materials it was enhancing the importance of others. There are no substitutes for minerals, and increased uses for such commodities as radium, nickel, mercury and tungsten have placed a premium upon their sources. Fortunately, perhaps, for the world at large no one nation has a monopoly of all these very necessary items of industry.

So far consideration of the influence of science on industry has been confined to the effects of the introduction of technological changes. Up to now this has been almost the sole extent of its participation, and great as have been the benefits conferred by this limited association there have been also grave drawbacks. Little or no control was exercised over industry ; a policy of *laissez-faire* was adopted both by Governments and by industry

---

[1] In South Carolina I have seen a large weaving shed without windows in which lighting and atmospheric conditions were completely artificially controlled.

itself, with the result that abuses of the power that technology had placed in its hands soon became apparent. The introduction of power-operated factories led to the congregation of workers in the industrial regions. The growth of the industrial towns of the north and midlands with their squalid surroundings and grimy atmospheres is one of the results of the intervention of technology. The nineteenth century saw the growth of an increasing grimness in the life of the workers ; long hours of labour, the employment of young children, and the spawning of slum property unequalled in the world. Happily much of this has since been rectified, though much remains to be done, but the avid utilisation of the sciences as uncritical servitors has been responsible for untold misery. To the credit of science we may add that some, at any rate, of the present-day improvement is also due to its intervention in the shape of increased knowledge of human requirements in the way of fresh air, food, leisure and exercise, and in the invention of new and easier or safer processes for carrying out previously unpleasant or dangerous operations.

Many of these evils were the outcome of the selfishness of men sufficiently powerful or sufficiently knowledgeable to employ scientific advances for their own ends. But this implies that such men were unacquainted with the discipline of science, and even though it may be admitted that great improvements have taken place, have we reason to believe that the modern industrialist is in this respect a more enlightened man than his predecessor ? He is probably more desirous of the technical help that science can give him. He is aware that dividends are paid in the research laboratory, but he looks as often as not for a quick turnover. Long-term, fundamental research is not usually his aim, although there are notable exceptions. Nevertheless, the full value of the contribution that science can make is not appreciated. The acceptance of the technical assistance of science has not been followed by any recognition that the methods of thought that produced the technical advances are equally applicable to every other problem of industry whether it be social, economic, financial, political or personal.

The leaders of industry to-day fall into three broad groups— men who have forced their way by means of their dominating personalities from the bottom to the top ; the large class of technicians in company procedure, accountants, lawyers, promoters and bankers ; and those who have inherited businesses or names valuable because of their association in business. Of these three groups, the first will almost certainly continue, for

it contributes something of real value through the qualities that
have enabled its members to rise, while the last may perhaps
gradually disappear from the industrial scene as names assume
their proper perspective. The leaders in the second group are
somewhat different. They contribute to industry's well-being
by virtue of the specialist knowledge that they possess. As
specialists they are useful, perhaps even indispensable, but it is
not necessary that they should be the sole arbiters of policy.
We should do well to remember that they are comparative
newcomers to the business world and that before 1862, when the
Companies Act which introduced the idea of limited liability
was passed, industries ran successfully without their aid. In
those days industries were run by technologists even though
their technology may have been founded on empiricism. What
it amounts to is that business and industry need to be run by
the appropriate technologists ; business by men who understand
and can use the tools of company law and finance ; industry by
men who understand its processes and appreciate the interrelations of its products. Commerce needs them both and it is a
lack of balance that has placed the ultimate control in one kind
of hands.

It is easy to see broadly how this has occurred. The inventions of the last century resulted in the laying down of industrial
plant on a previously inconceivable scale. At the same time the
growth of the joint stock banks with a conservative policy on
loans meant that business men were no longer able to raise
capital from a sympathetic local bank on the security of a good
name and an unblemished reputation. New inventions and processes were pouring out and capital was scarce. Men with the
capital to invest were able to make their own terms and to reap
handsome profits, while experts in the raising and handling of
capital and in the understanding of the many new safeguards
that were being instituted for business were in demand. That
day is now passing and a new set of circumstances may soon arise.
Competition is now so keen that inventions are attaining a
premium value and the men who understand and can work them
are in demand. It may well be that in the not far distant future,
industrial technologists will be able to dictate policy to financiers
who are anxious to invest money or to capitalists who are anxious
not to lose money already invested. The relative positions of
financial and technical understanding are slowly changing.

Nevertheless, although industry needs more technologists as
arbiters of policy, there is no guarantee that they will be more

successful than the financiers whom they displace. Both are simply possessed of specialised knowledge in a particular branch of learning, and gaps in the technologists' financial equipment may be as disastrous as gaps in the financiers' technical equipment. The truth seems to be that while both kinds of specialist are needed in the board room, they must, to be successful, be infused by one spirit. Technical skill, whether it be with high-pressure vessels or with double-entry book-keeping, is not enough. There must be a sanity of outlook, a balanced judgment in adversity as in affluence, an ability to correlate the broadest issues and to see the relevance and implications of widely separated facts of whatever nature ; a ranging, imaginative, disciplined mind. In short, the leaders of industry, whether financiers as they now are or technologists as they are becoming in increasing numbers, must also be scientists. Is there any reason why financial genius should be exempt from scientific training ? Is there any reason why the leaders of industry should not be deliberately taught to think positively instead of learning by bitter experience the pitfalls to avoid ? This is not applied science with a special leaning to particular fields, but what we are sometimes pleased to call "pure science," science for its own sake or, more properly, science for the sake of mental maturity. It is not scientists as experts that are needed, but scientists as mature individuals—not so much their professional knowledge as their ingrained method of thought.

This is the only way in which industry can solve the problems of unrest, of recurrent depression and of obsolescence, which have racked it for generations. It is one of the urgent necessities of our times. All are agreed that science holds the key to industrial salvation, but not all are agreed upon the way in which it shall be called upon to use the key, and the future of industry depends not only on science but upon its proper use. Sir Harry Lindsay voiced the general view recently when, with reference to the task of reconstruction after the present war, he wrote : "Now that war is upon us again, it behoves us to make wise and far-sighted plans for future development. The success of these plans depends on one essential factor, namely, the application of scientific methods. All economic development, if it is to succeed, must take careful account of the natural conditions underlying production, processing, transport and marketing."[1] The industrial success of Great Britain in the last century was made possible because she was first in the field, and she was first

[1] *Transition from Peace to War.* Bull. Imp. Inst., 1939, xxxvii, 513.

in the field because of the natural advantages she possessed in her vast resources of the coal and iron necessary for her own industrial development. These gave her a technical lead over her competitors, but the day of such natural advantages is past. Every nation has learned to exploit its natural resources and the hope of industry to-day lies in the technical equipment provided by scientific research and the ability to use it to the best advantage provided by bringing to the control of industry the same severe and disciplined thought that has produced its technical advances. Financial organisation and business reconstruction will not help in this race, for they are too easily duplicated in other countries so that their effects are exceedingly temporary and, moreover, they are not indefinitely progressive but come to an end in a relatively short time. This was recognised as long ago as 1922 by the Advisory Council of the Department of Scientific and Industrial Research, who recorded their opinion as follows : "It is well recognised that for four-fifths of their food and for a great part of the necessary raw and semi-manufactured materials for industry, the people of these islands are dependent on supplies from overseas. These supplies can be obtained only if this country is able to carry on its exporting industries in future with greater efficiency than the rest of the world, for it is doubtful whether we can compete, either by lowering wages beyond the limits of our competitors, or by securing a much greater human effort than they. If these two avenues are closed, competition, in the end, is confined to greater efficiency resulting from scientific work, for, in the long run, our outstanding business skill and organisation could not make good a deficiency of production or an obvious inferiority in our goods."[1]

Given the technicians of finance and technology in twin control and both informed by the spirit and method of science we may confidently look forward to something like a golden age in industry. Efficient production combined with decreased labour and ample leisure must follow in its wake. The flexible outlook that weighs change but is not afraid of it would permeate factory, mill, foundry and counting-house. Green belts, smokeless cities, sunlit homes would replace squalid slums, haze-hidden skies and grimy dwellings. The study and pressure of the social sciences might make for better relations between man and man. Economic geography, statistical studies of the world's wealth

[1] Seventh Annual Report of the Advisory Council for Scientific and Industrial Research, 1922, p. 11.

and needs and planned economy could curtail ruinous national and international competition. This may appear Utopian ; too idealistic ever to be realised. But who knows ? Changes are rapid in these days and given the opportunity, science could do these things.

One thing remains to be said. So much has been written concerning the benefits that science has conferred or could confer on industry that some may feel tempted to ask if the relation between them is entirely one-sided. Does science give all and expect nothing in exchange ? Or does industry accept all without making an adequate return ? By no means. On the technical side it has been a true marriage to which both partners have contributed and from which both have benefited. The greatest steps in scientific progress have often resulted from industrial needs which have caused scientists to delve deeply for explanations that have afterwards proved to have general applicability far beyond the confines of the immediate problem. The great Lavoisier developed his theory of combustion as a result of work on the lighting of the streets of Paris—the best lamp, the best reflector, the best oil-container and thus to the study of combustion. One need only mention further the work of Pasteur on fermentation and the diseases of wine which led to the spontaneous generation controversy and the study of infectious diseases. The research work carried out in industry for a particular purpose may be carried over into science to its immense broadening. The laboratories of great industrial concerns such as the Bell Telephone Company and General Electric Company in the United States, or Imperial Chemical Industries in Great Britain have contributed wide expanses of new knowledge to science.

The needs of industry have helped to keep the feet of scientists upon the ground even when their heads have been in the clouds. The philosopher's stone, the elixir of life and the phlogiston theory were the results of the segregation of scientists from the world of actuality. The greatest claim of Lavoisier to fame is based not upon his original discoveries but upon his sane and considered interpretation of well-known phenomena. "It was," as le Chatelier said, "his constant preoccupation with practical questions that enabled Lavoisier to escape without effort from the fictions and conventions amid which contemporary chemists were merely marking time." There may be some who fear that modern scientists are straying so far from practicalities that their theories are taking on a strange, unprovable metaphysical shape.

32   THE IMPACT AND VALUE OF SCIENCE

It is the influence of industry that will apply the corrective in the forward march of the sciences.

Lastly, the sciences have received great material benefits from industry. Industrialists have not hesitated to devote the wealth derived from industry to their furtherance. Research institutes of the type of the Rockefeller Institute for Medical Research and the Carnegie Institution in the United States of America, or various university laboratories in Great Britain like the Nuffield Physics Institute at Birmingham or the Wills Memorial Physical Laboratory in Bristol, are examples of this kind of industrial endowment. Many great industrial concerns also maintain fellowships or pay for expensive facilities at colleges, universities and research institutes which enable scientists to devote their whole time and attention to fundamental studies along lines that might otherwise be impossible. The debt that the sciences, and indeed all learning, owe to industry can hardly be over-estimated.

But all this interchange has taken place on the technical and material planes. The scientific method still passes unrecognised in the halls of industry and on the exchanges of commerce. It is by the adoption of its method that industry could best show its appreciation of scientific achievement and that science could make its greatest return to its material benefactors.

SCIENCE AND POLITICS

"Science walks hand in hand with human development as its constant benefactor, as the guardian of its peace, in a universe rich to provide happiness and security for all."

W. F. G. SWANN.

THE world's industrial face is being constantly transformed by the energies of technology and, by the transformation, new social and political problems are as constantly born to be faced, solved or side-stepped by succeeding administrations. The impact of industry upon politics is growing steadily ; not only is it responsible on the debit side for many of the most difficult problems posed to governments, but on the credit side it provides the sinews of financial stability. A successful industrial regime, whether it be primarily agricultural or manufacturing, is an essential to the well-being of nations. If there is any question as to the inter-play between industry and government it should be dispelled by even a cursory glance at a few of the great departments of State. They include the Board of Trade itself ; the Ministry of Labour, which now, among other activities, administers the Factory Acts ; the newly constituted Ministry of War Transport, with its vital sections of Railways, Highways, Port and Transit and Road Transport ; the Department of Agriculture and Fisheries, with such scientific sub-sections as the Ordnance Survey Department and the Agricultural Research Council ; the Post Office, with its widespread engineering commitments. They include too, in wartime at least, the great production departments—the Ministry of Supply and Ministry of Aircraft Production. They include the Board of Customs and Excise, dependent not only on domestic but on foreign industry, and ultimately they include the Treasury, utterly dependent on successful industry for the national revenue. A similar rapid glance over the titles of statutory commissions and committees that have been set up in modern times might be even more convincing. The preponderance of industrial and technical, not to say scientific, interests, represented by bodies such as the Coal Commission, the Electricity Commission, the Development Commission, the Wheat Commission, the Forestry Commission, the Imperial Communications Advisory Committee, is overwhelming.

The political scene thus controlled by industry must change with an ever-increasing acceleration as a result of the application of science in one or other of its many fields of endeavour. Acting through technology it forces changes, sometimes for the better and, alas, sometimes for the worse. The changes may be unforeseen but they are none the less inevitable. The American Association for the Advancement of Science recently issued a statement in the following terms : "Science and its applications are not only transforming the physical and mental environments of men, but are adding greatly to the complexities of the social, economic and political relations among them." Population changes, drifts from the land, increase or decrease in the standards of living, changes in the birth- and death-rates, imports, exports and their associated political features of protection and tariff, are all directly or indirectly traceable to the influence of the sciences. Our daily security is dependent on their progress. We owe as much to William Murdoch and the illuminating engineers who have followed him as to Robert Peel and the succession of Chief Constables. Oil lamps were first replaced by gas in Westminster in 1814. Before that time London, in common with other big cities, was but poorly lighted and crime flourished after dark.   Increased lighting has always been associated with decreased crime.

Is a new and cheaper source of a material found ?   An industry is ruined with its concomitant effects of worklessness, poverty, illness and increased public expenditure on assistance and relief. At the same time, in some other place, perhaps in some other country, an industry is created with its concomitant demands for labour and capital and a consequent drift of both from the old to the new.   The replacement of wrought iron by steel which could be produced at lower cost and possessed greater strength for the same weight was not only a scientific achievement.   It was a force which brought social influences of the first magnitude into play.   The puddlers in the iron works, skilled men who had spent a lifetime in the service of wrought iron, were rendered workless when puddling became obsolete.   Their families were reduced to the poverty line and they became dependent on their children or on poor relief.   This, in its turn, rebounded upon their sons, who were prevented from marrying or, if married, were left with so little that their children grew up under-nourished.   But at the same time new labour was needed for the steel works so that in the Welsh iron districts, for example, large numbers of English, Irish and Scotch were imported to serve the furnaces and their descendants are there still.

When the new material or the more efficient process is exploited, few, if any, take thought for the wider effects.   The nature of these secondary changes is not decided by science or by scientists, but, as a rule, by men to whom the advances of scientific thought represent opportunities for personal benefit or the advancement of individual policies without regard for other issues.   It is a procedure which permits technological improvements to wreak havoc rather than to spread good.   Yet the same mental equipment that gave birth to the technical advance could, if applied, ensure that it was used only for the improvement of the conditions of men.   Until this happens industry must continue to pass through periodic depressions and booms and the political results of such undulations must be expected.   Moreover, science for the same reason must pass through periods of severe distrust by laymen when it will be blamed for the ills and terrors that result from its indiscriminate employment.   Changes must come and science will ensure that they do so more and more rapidly for it is the modern driving force behind the transformations of society.

There is, in the steady enlightenment of the public conscience, a directing influence to maintain that the changes in life and culture wrought by scientific endeavour shall flow in channels of usefulness and goodness. But direction without power is useless. It would be a beautiful dream to plan the emancipation of men from the enthralment of industrial circumstance, but it would remain no more than a dream if science by its invention did not make it possible. Behind every modern major political issue stands the commanding figure of science ready to be used for good or ill; impersonal and unmoved by the results of her discoveries, but willing to determine that they shall be for the benefit of mankind. Acting through technology science enforces change. The same science acting through social consciousness could ensure that the changes for the better far outweighed in number and importance those for the worse. It is an indictment of our present system that the power for evil given by science to unscrupulous men seems so far to outweigh the power it confers to do good.

We must, therefore, consider the relation of science to politics, which may prove a difficult proceeding, for politics affects the most intimate details of daily life of scientists as well as laymen. It is inevitable that, being men as well as scientists, they will hold strong views on political matters, views that will not necessarily be the product of scientific thought. In politics the would-be objective investigator finds himself both sitting on the bench and standing in the dock. Any decision that he takes will have some sort of, possibly some very direct, repercussion upon himself. The scientist in his laboratory, apart from the system he is studying, can look at it with all the objectivity in which science has trained him. When he leaves his laboratory for the hurly-burly of life he finds this limitless objectivity difficult of attainment, but even so he should be able to establish a fair degree of critical and unbiased judgment for conceivably, in industry, economics or social science, he may be able to employ his peculiar ability even though there may be hardly a question to be decided that will not, according to the decision taken, affect himself for good or ill. The new system of thought must be applicable to politics, not as readily perhaps, but every bit as surely as to other more amenable branches of human effort.

Obviously, before any such application can be attempted it will be necessary to define the aims of government. In the dictator countries the aim of government is unequivocally stated to be power—the unlimited control of men and materials—and the behaviour of their governments conforms logically to this view. In consequence, civil rights are steadily lessened, free speech is denied, criticism is stifled and resistance to the regime is rigorously stamped out. Such a clear-cut picture cannot be given of a democracy, in which the ideal of government lies in service to the community. The degree in which this ideal is attained varies with the country and with the times. Unfortunately, the democratic ideal with its roots in the past and its peculiarly misty horizons is particularly susceptible to half-disguised dictatorships masquerading under well-known and commonly accepted pseudonyms, and the absence of scientifically trained minds from the nations' councils makes it easy for perfectly honest men to deceive themselves and others as to their aims. Science and freedom of thought go together. Democracy and science sprang from the same soil and grew up together, and the lesson of the dictator countries is that they will die together. Although it may be impossible to predict what a scientifically trained government would be like, its policy should at least be patently obvious from the first, for it is a prerequisite of scientific thinking that the problem should be clearly stated and commonly understood.

But while science might provide men capable of realising the democratic ideal, "No society," as Hogben says, "is safe in the hands of a few clever people, without intelligent co-operation and understanding from the average man and woman." It is not sufficient to have scientific leaders. We are dealing with an outlook, a point of view, a system of thought that becomes more powerful as it is more widely adopted. It is not, by its nature, capable of being imprinted upon all and sundry by the exercise of force. Intrinsically it depends for its strength upon understanding and free acceptance. The scientifically trained mind, imbued with the ideal of freedom of thought, grounded in the system of unbiased judgment, might arrive after much weighing of the available evidence at a policy to pursue but it could appeal only to similar minds for understanding and support.

Unless a nation possesses a substantial leavening of independent minds it is doomed in fact, if not in evident practice, to a dictatorship of some sort. So long as men can be swayed more by their passions than by the evidence presented, so long will they be susceptible to false arguments. This is not to condemn passion. A righteous indignation against substantiated wrong is something of which to be proud, but these same passions can be, and often are, roused to fever heat by catchwords, by appeals to base emotions of envy, jealousy and pride and by high-sounding but ill-founded arguments. Democracy of such sort is no more democracy. It becomes mob-law. But men, even in mobs, are not necessarily vicious. It is ignorance in the main that permits support for wrong, and for ignorance there is only one cure—clear, unequivocal, forceful statement of truth presented to minds trained to accept it. "In a democracy there should be individuals in all walks of life who have been trained to think clearly and without prejudice, who are able to perceive intricate inter-relationships between problems."[1] The more we have such people the greater will be our distrust of mob-law and the more powerful our weapons against it.

In this connection one cannot escape the conviction that the extensions of the franchise in response to popular appeal, though seemingly truly democratic, were in fact retrograde steps which have debased the democratic ideal. They have permitted the incursion into politics of the loudest voice and the most plausible tongue rather than the deepest thought. What was needed was a restriction of the franchise to a capable few—say to a certain level of education—and then the widening of that class by increasing educational facilities until the whole nation came within it.

But to return to the leaders. If it is desirable that there should be trained and independent minds among the rank and file in a democracy, how much more important it is that the government of the country should be in the hands of men trained to think clearly and to weigh evidence. It seems pertinent to ask what sort of government it would be. The answer to such a question can be given only in the most general terms and it is no part of the present task to try to outline the policy that a scientifically minded government would develop. This would entail far too deep a study of present-day needs and an analysis that could be undertaken efficiently in fact only by the kind of government that we are envisaging, though it may be pointed out that organisations already exist for the objective study of political needs and economic problems.

There is, however, every reason to believe that a government of scientists might be a most truly democratic government. The burning passion behind all scientific investigation is improvement, and if scientific thought were brought to bear upon the problems of the nation it would be the betterment of the nation,

[1] "The Road Through Adolescence." Prof. Olive Wheeler in *Educating or Democracy.*

the improvement of international relations, the welfare of individuals that would decide policy. All scientific work ultimately subserves the needs of humanity and there is no conceivable reason why scientific government should prove an exception. Policies under a scientific government would be the result of constructive national and international planning based on secure knowledge and sound premises rather than on the exploitation of causes of friction designed to benefit the few even though it led to the despoilment of the many. This could not be brought about at once. The science of government is one of the most imperfectly understood. The study of human relations stands to-day where the natural sciences stood before Newton. The truth is coloured by what men want to believe and by opportunism rather than by incontrovertible fact, and the entry of the spirit of scientific enquiry into the realm of policy is feared for the truths it might discover, for the shibboleths it might unmask, for the feet of clay it might reveal.

Enquiring minds are viewed with suspicion. They challenge established beliefs at regular intervals and so make for discomfort and an uneasy awareness of the mutability of human institutions. Among themselves scientists are used to propound hypotheses which must run the gamut of criticism, and those that live pass a statistical survival among all sorts of minds. Scientists are accustomed to wait for recognition. They are not politicians whose words spoken to-day are gospel to-morrow and forgotten the day after. It takes time for the acceptance of scientific advances. It was thirty-five years before Newton's *Principia* was taught in his own University, and Mendel, who published his theory in 1865, died in 1884 a disappointed man whose work had gone unrecognised. Even when theories have been commonly accepted they are challenged by each new generation in the light of fresh knowledge, and they continue, are modified or fall according as they meet the scrutiny. Inquisitive scientists have the awkward characteristic of seeing no good reason why established policies and pet theses should escape such periodic bombardment. In fact, there is no good reason ; there is no reason at all save either the venality or the inability of the subscribers to the shibboleths. "The truth is, sir, that the institutions of men grow old like men themselves, and, like women, are always the last to perceive their own decay."[1]   For this reason the light of disciplined minds must not shine too brightly on politics lest

[1] "Letters from England." Don Manuel Espriella. *Edin. Rev.*, 1808. Vol. XI, p. 378.

they reveal the old harridan for what she is. The truth that scientific enquiry would insist on unearthing might be offensive, and it is better that the perfumes and powders, the cosmetics of careful dissimulation should hide the ravages of time and the evidences of evil living.

It seems likely that the return of scientifically minded men to power would be conducted in a different fashion from what we now know. The specious promises and Party slogans of to-day, with their half-truths or utter falsehoods, would be things of the past. Indeed, it is possible that Party government itself might be swept away. It is difficult to imagine how thinking men could at the same time be good Party men. The Parties, after all, have a definite programme for specific groups and were formed to safeguard the interests of these groups. But the population of the country is not all working class, any more than it is all upper class or all agricultural class, and although any Party may on occasion have desirable programmes, and although they may change with time (despite their names), yet independent minds unafraid of solitary thought could hardly give whole-hearted and uncritical support to a Party programme. But that is the rule to-day; and more, it is not even necessary that the candidate shall give his support whole-heartedly. All that is necessary is that he shall cease to think for himself. The Whips' office will think for him and tell him how to vote. In a crucial division where opinion against a measure is sufficient to make its passage doubtful he is not even allowed the privilege of abstaining from voting. It was, I believe, the Member for Derby who abstained from voting after the Munich crisis and who was subsequently called upon by his local association to explain why he should not resign, not because he had opposed the Government but because he had not actively supported it. In a scientific government there could be no idols of wood before whom to bow down and worship. A Member's convictions of the well-being of the country would be of greater consequence than Party affiliations.

There are occasions when the Whips are removed and members are allowed to follow the dictates of their inclination. Such an occasion occurred recently when a Bill was introduced to provide for pensions for Members of Parliament. Sir John (now Lord) Simon introduced the Bill and the report added the significant remark that it would not be a government measure but that votes on it would be free and the freedom to oppose it would extend to Ministers. It is inconceivable that a democracy would accept such a statement with its natural implication that votes are not normally free, that Members are not present to represent the wishes of their constituents nor the dictates of their own considerations. Jack Jones, in his book *Unfinished Journey*, records that as a Member of Parliament he was bombarded with instructions from the Party "until at last I ignored the instructions and carried on as I thought best for those whose representative I was." It is unlikely that trained minds would consent to be thus fettered by a Party, dragooned by a Whip. This perhaps accounts for the scarcity of scientists as Members of Parliament. Few of them have leisure or financial backing to stand as Independent candidates and they could hardly suffer dictation without losing their hard-won freedom of thought. At the Cambridge University by-election in 1940, the candidates were each asked if they would take an ordinary Party Whip or a Whip "for information."[1] From their replies the questioner was pleased to note that whichever was elected the University would be assured of independence in its Member. There seems to be a consensus of opinion that University Members should preserve an independence of outlook. In what way does a University Member differ from the Members for other constituencies ? Is it a cynical recognition that a reputedly intelligent electorate who must be won by the coldly written word rather than the passionate platform plea is less likely to be humbugged than the patient electors of a normal constituency ?

Government by a cabal, which is what Party government with its regimented thought in the rank and file has resolved itself into, although not truly democratic is not necessarily a bad thing, for the cabal may be composed of enlightened men with the good of the nation at heart. When that inner group, however, have affiliations other than their common love of truth and the welfare of their people, then policy is dictated by considerations other than national necessity. In recent years we have seen too often how when governmental policy has been so dictated normal standards of honesty are no longer considered to be necessary in Parliamentary affairs. Expediency and self-interest have now become surer guides to the actions of Governments than honesty of purpose and uprightness of character.

Prejudice, jealousy and intolerance are the natural offspring of political parties. Associations of whatever sort induce in men a vicious loyalty which makes them aggressive and vindictive and sometimes leads them, in the interests of the Party,

[1] F. P. Salter in the *Daily Telegraph*, February 21st, 1940.

to tamper with the truth. The evil characteristics that manifest themselves in men—in well-meaning men—whenever they segregate themselves into groups, whether they be political parties, religious sects or philosophical schools of thought, are an incalculable hindrance to progress. The lesser object becomes the supreme motive to which all decisions must be subordinated. This shows itself in politics by disregard of the public conscience and suppression of unpalatable truths. No government is free from these characteristics. *The Observer* recorded shortly before the present war that there was "a new note of fearless objectivity in the British establishment of fact."[1] It is true that before that time the Government with fixed ideas had resolutely closed its mind to every evidence contrary to its adopted policy and nothing short of imminent catastrophe could shake it into honest objectivity of outlook. Who knows but that such fearless objectivity adopted consistently and not alone in the face of danger might have averted the catastrophe. The ties of partisanship, however, have been too strong and the scientific outlook has been sadly lacking in British as in other governments for many years.

Whether or no we have a scientific government it seems obvious that the governments of to-day will inevitably give way before a more intelligent system. It is probable that the political machine never was very efficient. In many respects it has been designed, or rather, since design has too great implications to ascribe it to politics, it has grown up to permit, even to foster, obstructionist tactics, and like many another piece of machinery, so long as it performed its work to the satisfaction of its owners and showed reasonable returns, efficiency was not a prime consideration. Unfortunately, like other machinery too, the older it becomes without serious overhaul or replacement, the less efficient it becomes until ultimately in a changing and competitive world it may be a sufficient drain on its owners' resources to put them out of business.

It is, then, a question rather of external forces than of internal dissatisfaction with the machine that dictates change. What are the particular external forces that are likely to enforce changes in politics ? The greatest probably is the increase in the speed and efficiency of transport and communications, which is a purely scientific achievement. It started when Stephenson's Rocket drew its first load from Stockton to Darlington and spelt the end of the stage coach ; when the *Great Eastern* first churned

---

[1] "The World Week by Week." *The Observer*, July 2nd, 1939.

C

its laborious way almost independently of the winds across the Atlantic; when the internal combustion engine was invented and so paved the way for the brothers Wright to risk their lives at Kittyhawk by actually flying a heavier than air machine; when Bell installed his first telephone and when Hertz made his experiments on transmission of sound without wires and so opened up the path that led Marconi to the wild dream of setting up a transmitting station in Cornwall to communicate with Newfoundland. Politics alone has so far failed to recognise the implications of the new era. Science, art, music and all learning, communications and even competitive industry are now international in character. Only politics stands aloof preserving the attitude of pre-communication days, enfolding itself majestically in the gossamer of sovereign power.

Before these events the north of England was hardly affected by occurrences in the south, but now a speech at Westminster has undreamt of repercussions in Washington, Tokyo, Bombay and Buenos Aires within the hour. This means that whereas well-meaning but incompetent men were once able to perform the functions of government for years without serious mishap by a strict conformity to precedent, the same men are now a source of national danger. The method of procedure by precedent is expressly designed to suit the mentality of dull men who are unable and have no desire to think deeply. But the dilatory political methods of the 1860's are inappropriate to the tempo of the 1940's. It is as if the Rocket were harnessed to the Silver Link or the paddles of the *Great Eastern* to the *Queen Mary* to carry her across the Atlantic in search of the Blue Riband. It is no longer enough that a man should be a good fellow and a staunch party adherent. It is not sufficient qualification for high office that a man should have given many years of humdrum, faithful and uncritical service to an outworn creed. By all means let him be rewarded for such, but rather by relieving him of responsibility than by thrusting it upon him.

Are we, then, to advocate a government of scientists? Why not? As men they are not likely to be less efficient than others, and as scientists they may be considerably more so. There is an exhilarating attraction about the idea of a new ruling class of scientists speaking no special language of the rich or poor, the leisured or the working classes; exhibiting the scientific ideals of tolerance, justice, impartiality and altruism. There would be some mighty changes as "the vague and foggy generalities that so often pass for a wide humanistic outlook" were subjected to the searching exactitude of the scientific method. Mists that have befogged the minds of men for generations would be swept away ; miasmas that have poisoned their relations would be dissipated. The powers that the sciences have brought within our reach would be harnessed for the common good by men who understood them. It paints a rosy picture. Could science do all this ? With one provision—yes. Science makes no claim to distinguish right from wrong. It claims only, by permitting men to envisage the logical and inevitable outcome of their courses, to make it possible for them to pursue the good successfully if they are so minded. Whether a government of professional scientists could achieve this end is more difficult to answer, for as D. L. Watson says : "Scientists are Human."[1]

It might be preferable, however, to think of a government of independents, for it is not a government of expert scientists that we want, although it is certain that there would even to-day be a fair sprinkling of scientists in such a gathering. They would be intent on plumbing the fundamental causes of national unrest and international friction. There would be differences of opinion even among the leaders, but these would be no more than swirling eddies on the surface of a stream constantly and relentlessly pursuing its appointed course to the wide ocean of men's best desires. But swirling eddies may sometimes be magnified to the turbulence of rapids, menacing and dangerous, and broad streams may overflow their banks with devastating effects. Who, among a government of independents, is to maintain the steady direction of the political stream ? From whence are the leaders to come if they have not the assured support of a predominant Party sworn to allegiance ? The deprecating smile is already forming on the face of Mr. Worldly Wiseman as he contemplates even the initial difficulties of so half-baked a suggestion. But how are leaders found in a Party, whose members though linked by a common aim, remain among themselves a heterogeneous crew ? And so, even among independents, there will undoubtedly be men of outstanding ability to whom the remainder will look for guidance. True, they may be defeated in debate on occasions, for they will not have a regimented majority to depend upon. But it will at any rate be in real debate, for without a dependable and amenable backing every motion will have to be presented on its merits, a real case prepared, and sincere opposition fought. The House of Commons scene might be revivified as the kaleidoscope of argument

[1] The title of a book published in 1938.

shifted and to-day's opponents became to-morrow's supporters. It is a pipe dream only, of course, but not so far removed from reality as might appear.   To win this war Tories and Socialists will sit in common council with Liberals, each preserving his own point of view but combining to achieve a single end because victory is so well worth winning and this is the way to ensure it. And if one aim can be welded out of the furtherance of the war effort, why, in peace, should not the benefit of the nation and of mankind be forged into a single all-consuming loyalty before which lesser loyalties recede and independence of thought becomes once more a cherished possession.

SCIENCE AND WAR

"Science, whatever harm it may cause by the way, is capable of bringing mankind ultimately into a far happier condition than any that has been known in the past."

BERTRAND RUSSELL.

No discussion of the relation of science to the activities of men would be complete in the century in which we live without some reference to war.   Before the half century is reached the whole world has been bathed in blood.   A world-war, the greatest of all time, has been fought with all the resultant devastation, horror, misery and degradation that that implies.   Millions of men, knowing little or nothing of the cause of the quarrel, have been locked for more than four years in a deadly embrace.   As a result, so we are told, the world has been purged, the dross has been removed as by a purifying flame, men have been ennobled by the endurance demanded of them during those weary years, and now the world starts on a new course.   The war to end wars has been fought.   Yet, in this new and perfect world, for which millions died that it might be born, we hear more constantly than ever the distant rumble of gun-fire and the rattling of swords only half-disguised as ploughshares.   In the deadly swamps of South America we hear it, on the dust-covered plains of China, amid the hills of Abyssinia, under the Mediterranean sky of Spain and now at last, once more, from an incendiary dropped in Poland, full-scale, world-wide warfare has burst into flame.

All are agreed that war is one of the most futile of man's

insanities and yet it goes on. Ever and anon, the fever spreads with epidemic vigour and men lose themselves in a fervour of hatreds unmatched in the animal kingdom. Only the rats, I believe, exhibit anything approaching the cannibal destructiveness of man at war. They alone, the loathsome, horrid creatures to whom we devote a special week each year for their extermination, prey on one another and organise destructive warfare against their own species in human fashion. But even the rats have a lot to learn from the lords of creation, for the genius of science has transformed war ; transformed it so completely that it almost looks as if war may be not only man's greatest folly but his last.

As in all the activities of men, science can play a double role. We have already pointed out that science does not concern itself with right or wrong, indeed that it recognises no difference between them. Science merely says, if you do so and so the result will be so and so, and this will be true whether what you do is right or wrong. It is for you to choose the course, but having done so the end will be certain. Hence, if we decide that war is either desirable or inevitable science will help us to wage it successfully. On the other hand, if we have not yet decided that war is inevitable, science will show us whether our behaviour will end in war and what behaviour we should adopt to avoid it. We seem, in the twentieth century, to have decided that war is inevitable, and the sciences, accepting the decision, can be pressed into service to make the war efficient.

In the present circumstances, scientific technology is vital before, during and after any major war. A nation unprepared by years of scientific research dare not go to war. If it has not the facilities for the researches necessary within its own frontiers then it must pay for the results of scientific research conducted in other more favourably placed countries. No nation is exempt from this compulsion and none has failed to respond to the need to go one better than its neighbour. Gatling, Hotchkiss and Maxim, inventors of machine-guns, were all Americans. Nordenfeldt, another inventor of a machine-gun, was a Swedish engineer. The submarine was invented by the French ; the tank is due to the British. Nobel, a Swede, was the greatest explosives manufacturer the world has seen. Lewisite, one of the most effective of war gases, receives its name from W. L. Lewis, an American professor of chemistry, while ethyl-iodo-acetate, which may be used as a lachrimator, is known in this country as K.S.K., after the initials of South Kensington where it was prepared. Haber,

whose process for the manufacture of synthetic ammonia from
the nitrogen of the air probably saved Germany from premature
defeat in the last war, was in supreme control of her chemical
warfare department. Each nation has its secret weapon that in
the imagination of ignorant politicians is going to wipe the enemy
from the land, the sea or the air and so gives them the courage
to go to war. Fortunately for sanity, science not only portrays
the course of actions with certainty, it also portrays its own
limitations long before the limits are reached, so that, with a
little forethought, scientists can, if they are trusted, say what
new and deadly weapons are likely to be available and to what
extent improvements in present armaments can be expected.

It is not alone in armaments that science is called upon to
play a major role. The standards of living which science has
made possible in every country cannot be lowered at a moment's
notice because the country is at war. Means must be found to
maintain them during wartime and only science can do this.
No political speeches or appeals to patriotism will suffice ; hard
scientifically established truths are the only things that count.
So a nation must make itself self-sufficient. Science is willing
and it can and does do what is asked of it—but without consideration for the consequent upset in economic relations and world
trade. Science could have told us that these things would follow
and that they would themselves lead ultimately to war, but it
is not this side of scientific clarity of thought that is wanted.
We must exploit our own resources, at an economic wastage rate
if necessary, in order to be independent of all the nations who
might be able to interfere with our supplies in the event of the
war we are thereby hastening.

It has already been pointed out that the great alkali industry
based on common salt was the outcome of the fear that overseas
supplies would be cut off by a foreign blockade. Similarly,
margarine now produced in colossal quantities to provide a
substitute for butter was introduced in 1870 as a result of a prize
offered by the French Government for a successful butter
substitute. The prize was won by the French chemist Méges
Mouriés, who prepared a so-called "Oleo margarine" by exposing
beef fat in alkaline solution to natural pancreatic juices in the
belief that it was some such natural process that produced
butter fat. It is not for nothing that the German people have
become acquainted with the term "ersatz." The people must
have vitamins, even though they are urged to prefer "guns to
butter," transport must have petrol even though the world's

oil-producing centres may be closed to us. This is the role we permit science to play and she plays it well. She might be happier using her resources for the betterment of mankind, but we have decided otherwise. So the last war dragged on for more than four long years because the German scientists were so efficient, and the war of 1939 will be determined by the relative calibre of the activities of unknown scientific laboratories in the belligerent countries.

An excellent example of the power wielded behind the scenes by unknown and unsung scientists is afforded by the rapidity with which a method of combating the German magnetic mine was devised by Great Britain. The degaussing girdle was devised and developed at one of H.M. naval establishments with the advice and assistance of scientific men who were consulted for the purpose. The part played by the scientists was recognised by Mr. Winston Churchill when introducing the Navy Estimates in the House of Commons on February 27th, 1940. "We see our way," he said, "to mastering the magnetic mine and other variants of the same idea. How this has been done is a detective story written in a language of its own . . . we do not feel at all outdone in science in this country by the Nazis."

If science is necessary in the leisurely days of peace to prepare for an eventual war, how much more is its importance enhanced in the fevered rush of war-time emergency. Previously unsuspected difficulties occur, unexpected shortages appear and solutions and substitutes must be found at a moment's notice. It is unnecessary to stress this point. Warfare makes unprecedented demands on men and materials and these demands must be met at an unprecedented rate.

It is not so acceptable that science will be even more necessary at the conclusion of hostilities. This surely is casting it for a major role, and that we are not prepared to do. Science is regarded in much the same light as the soldier, to be trained before the war, to be encouraged and lauded during it, and to be relegated to the background with as little fuss as possible after it. We can do that with science if we wish. There is nothing and nobody to stop us. But let us bear in mind that science has propounded laws of the conservation of matter and of energy, and teaches that we cannot create from nothing. If we spend millions a day to blast, explode and utterly destroy the work of millions of men, there must eventually come a day of reckoning. No good then to bleat about reparations, goodwill, or a world fit for heroes to live in. What is needed is scientific investigation on a gigantic scale to provide for the material needs of other-wise hopeless men and women. It is science that wins a modern war, but it is no less science that will win the ensuing peace.

Because of its increasing importance in warfare there is a growing disposition to blame science for war itself—or at least for its modern horrors. But has science so increased the horrors of war ? It has widened its scope. The instruments of destruc-tion are capable of carrying their messages of death to greater distances and to vaster multitudes than hitherto, but so far as individuals go it is difficult to see that modern carnage is any worse than in previous ages. The arrow in King Harold's eye at Hastings was just as painful and every bit as fatal as a modern Bren gun bullet ; so too was the molten lead and boiling pitch poured from the battlements upon the besieging ranks beneath. The cutlass of Drake's day could inflict a nasty wound and the bayonet was a favoured weapon before warfare had taken up the inventions of science, and even now "cold steel" is still the most feared. Moreover, the same sciences that have produced at man's behest the high explosive, the poison gas, the bombing plane and the submarine, have made a repetition of the Crimea impossible. The fatal wounds of yesterday are the superficial scratches of to-day. Drugs and dressings, antiseptics and ordered public health prevent the wastage of life by infection and disease that was the bane of all previous armies.

It is not science but the use that is made of it that determines its good or evil effects in the world. The steel cylinders that bring nitrous oxide for the relief of pain, or life-giving oxygen in hospital wards, can be used to carry chlorine gas in wartime. The machinery that cleans and bleaches cotton for surgical dressings does the same for linters intended for the manufacture of gun cotton. The advances that science has directed in agri-culture are good, but when in unscrupulous hands the increased yield is cornered then harm results. Science is not necessarily to blame for the harm that follows in its train.

Let us be quite clear on another point. Science has no responsibility for war. The responsibility for that rests on every citizen. What the sciences have done is to transform it, but it existed long before they were thought of. It is the outcome of men's passions, their greed, hatred and cruelty. Science may be the assistant, but it cannot be the cause of war. It has not changed its essential nature, but by its inventions it has brought to every human being the knowledge of war's cost, waste and futility. The aeroplane, the wireless, the printing press which were so praised when they disseminated peaceful news are now the hated instruments which bring war to our doors. It is not war that we hate and fear, but our own proximity to it. So long as it could be confined to small regular armies we saw no reason to abhor it as an instrument of policy. So long as it was confined to China or Spain we were content to be platitudinously self-righteous. It is when science, doing its best for us who have already decided in war's favour, makes certain that none of us escape its effects that the world begins to censure, not itself for permitting war, but science for bringing it so unpleasantly close.

Then where does the responsibility rest ? Have scientists no responsibility for their discoveries and the uses to which they are put ? They have the responsibility of every citizen, no more and no less. Scientists, by the very nature of their calling, are reasonable men and reasonable men hate war. The truth is that knowledge has outrun wisdom and the power that science confers falls only too often into non-scientific hands. We are all responsible for allowing this to happen. There is no machinery to enable scientific men to control scientific discoveries. Governments which are unscientific can have no conception of the forces they play with so lightly. With their limited vision they can wreak havoc on unsuspecting millions. In war as in peace it is not science, but the unscientific hands into which its power falls that must bear the responsibility. But surely science and scientists could stop war. Yes, given the power and the confidence of nations. But if scientists refused to have their science prostituted to evil ends, would that not stop war ? No ! Wars would continue to be fought without the aids to their efficient waging that science has provided. *Modern* war, specialised and departmentalised as it is, might conceivably be prevented by the refusal of scientists to collaborate, but this is true of any group. Financiers could stop modern wars by refusing to lend money. Miners could stop wars by refusing to hew coal. Engineers could stop wars, railwaymen could stop wars, doctors and ministers of religion could stop wars.

Almost any nation, if so minded, could put an end to modern war. Science has put the means of stopping it into the hands of every first-class nation and some small ones. War has become so complicated that innumerable raw materials are necessary and they do not all occur in one spot. A major war cannot be carried on without petroleum, high-grade alloy steels, nickel, rubber, aluminium. These are in the control of nations.

The importance attached by governments to self-sufficiency

has already been emphasised. Some success has been attained with petroleum and rubber, but it is difficult to *ersatz* minerals. Chemists are clever and sometimes brilliant, but they are not miracle workers. Not so long ago the subject of sanctions was much in the public mind. Collective security could be reached at a small cost by the universal application of sanctions, not applied wholesale to the destruction of the economic life of a nation and the undermining of the spirit and morale of its people, but applied solely to the necessities of war—the copper, tungsten, nickel, aluminium and possibly rubber and petroleum. The common people would not feel such an imposition, but their leaders would realise that war was out of the question and might be tempted to see reason. In this respect also scientists bear the responsibility as men that all citizens bear. We cannot condemn and get rid of science, which has placed such power in our hands. Rather must we condemn and get rid of the unscrupulous men who have the power to exploit it for their own ends. At the beginning of the present war there was great talk and great fear of bacterial warfare. Was it from scientists that we feared it ? Bacteriology has done nothing but good for men. No, the danger came from political leaders into whose unscientific hands was concentrated by proxy all the power and knowledge that science has accumulated but without the wisdom for which it has so long striven.

Then it appears as if wars must continue and science must continue to play its part along with all the otherwise peaceful pursuits of men. Not necessarily. All that science has done for the "improvement" of war could have been done for peace. There is no outstanding problem between nations that the methods of science could not solve at a hundredth of the cost of a major war. Science itself is international in scope and outlook. Just as it knows nothing of good and evil, so it fails to recognise frontiers. Because of this it is unpopular. It cannot take advantage of the differences between nations, to exploit them and increase them. It can study them and bring to light previously unsuspected harmonies. In so doing and in the clear-sighted pursuit of peace it may demand unpleasant sacrifices and the greatest from the most favoured nations, our own among them. There can be no doubt that many men in every country, some of them statesmen or politicians, are anxious to ensure peace, but being unable to see clearly the outcome of their activities they are incapable of performing what they most desire. It is not sufficient to have good intentions. More harm has often been done by incompetent men of goodwill than by clever rascals. Goodwill there must be, but it is necessary also to be able to think clearly and logically and to envisage the ultimate ends of policy before assurance can be reached.

Nations will agree while it suits them to any policy. Germany was glad that France was building the Maginot Line in 1934. Japan resented, but agreed to the 5 : 5 : 3 ratio of naval construction, but soon changed it in her favour when the time was appropriate. Treaties and agreements are not sufficient to prevent war nor to reduce its horrors. War is not a huge game of football to be played according to rule, but a life and death struggle in which the losing side will stop at nothing to escape defeat. For example, chemical warfare, which is defined as the use of any substance in the realm of chemistry, harmful to the human or animal organism (*but excluding recognised high explosives*), is prohibited by the Geneva Protocol of 1925, but it is common knowledge that intensive research into so-called chemical defence was being prosecuted in every country of importance and the first reaction of Great Britain to the threat of war was the issue of gas masks to civilians. No amount of legislation for the reduction of this or that form of armaments will avail either to prevent or to minimise warfare. On the other hand, was there ever a more ridiculous theory than the one so frequently propounded in recent years that the road to peace lies through intensive armament so that no one dare attack. In the days before law and order was established in the wild west of America every man carried a gun and was ambitious to be "quick on the draw," for his life might depend on it, yet no man was safe. The march of civilisation, however, did not mean that every man had bigger and better guns, but that, for the safety of all, all were forbidden to carry guns on pain of severe penalties, and the preservation of order was entrusted to a small section of the community acting under discipline and according to a recognised law. The moral for nations is clear.

Before we can hope to prevent war we must know the underlying cause of war. It is here that the method of science by impartial examination of the facts could help us. It will mean painstaking and objective study for the apparent causes may not be the real causes. The international situation in 1939 bore a strange similarity to that in 1914. The iniquities of the Treaty of Versailles were being loudly proclaimed as the cause of war. It may have been the excuse, but it was no cause. The present situation and that in 1914 are the results of similar, perhaps identical, causes which started a long time ago and they will still exist when the present war is over, unless men and women are prepared to consider fundamental causes without rancour, prejudice or selfishness. No amount of bombing, fighting or naval victories will provide a solution of the problem. Until we know the cause we cannot hope to find the remedy. The murder of an Archduke at Sarajevo may precipitate unlimited slaughter, but for all his noble blood nobody is likely to believe that he was worth a world war or that he was the real cause of it.

That being the case it behoves us to be very sure for what we are fighting and whether, indeed, there is anything worth fighting for. It was simple in the days of old. The bold barons had something to gain by victory. Their lands and lieges were increased at comparatively little expense and their rivals were destroyed. Even so late as Waterloo, and perhaps even to the Franco-Prussian or Boer Wars, the rewards were not incommensurate with the cost in men and materials, but the power that science has conferred on governments has permitted war to get out of hand, to become so universal, so costly, so destructive, that it might be argued that there are now no positive results that can be said to be worth it. At this point someone is sure to ask whether honour and right are not worth fighting for, and the answer must be that circumstances can arise between nations as between individuals when, at whatever cost, a stand must be called in defence of recognised principles. But we must be careful to ensure that the cause is truly a righteous one, for every war is to both antagonists a righteous war, if only a self-righteous one. And when we have decided scientifically and unsentimentally that our cause is just we must see, with the same realism, that the fruits of victory do not elude us. In the last war we fought for the sanctity of a "scrap of paper," with the result that treaties are now not worth the paper they are written on and are universally regarded as liabilities rather than safeguards ; and for lasting peace and the security of small nations so that Abyssinia, Czechoslovakia, Poland, Finland and Albania could be safe under the care of the Great Powers. The men who fought and died in the last war believed in these ideals. It is we who live who have broken faith. And the young men and women of to-day believe that our cause is just and they, too, are fighting and dying that a better world may be born.

We must be certain, too, that in the times of peace which lie ahead we are not through cowardice drifting into war on the grand scale, and when war comes, if it must, we must be equally convinced that it could not have been averted by competent leaders. War, like almost every other political manœuvre, is an expedient. It is designed to prevent a possible immediate damage without necessarily making provision for a future welfare. It is a surgical operation of the severest kind performed to preserve the health of nations, and though we may on occasion face with fortitude and determination the prospect of drastic surgery we cannot be expected to be grateful to, nor to put our faith in, doctors whose proved incompetence or negligence made the operation unavoidable. The health of nations, like that of individuals, depends largely on wise living over long periods and not on a series of crises drastically surmounted. Science can provide us with the method of thought and the honesty of conviction that will secure peace if we desire it, and with the technological advance to maintain it.

SCIENCE AND EDUCATION

"Free intelligence as such has an elasticity of its own. The mind in its spring puts itself forth on all sides. It requires no stimulation, but only to be directed. The reason by its own nature, seeks truth. The young mind desires to know, to explore the unknown, to find out the nature and causes of things."

MARK PATTISON.

IF there is any truth at all in what has been said in earlier chapters then the relation between science and education assumes a great importance, for it is through the medium of education that the new method of thought will have to be inculcated. It is not possible to superimpose a new system upon adults, whose thought processes have already been moulded. Too many people, however, have already postulated education as a cure for the evils that have befallen the world, and while it is easy to accept this as a general hypothesis it is not easy to obtain agreement upon the precise educational structure that will best meet the need. It is widely argued that topical affairs should receive a prominent place in school training; or that education in citizenship should be a specific and important item in the curriculum; or again that a broad international outlook should be fostered by greater attention to foreign languages or geography. All these are doubtless useful suggestions, but, without wishing

to add one more to the already swollen ranks of amateur educational philosophers, it is desirable to enquire whether any or
even all of them are more than facets of a greater problem.
Surely what is needed in education, as in other walks of life, is
something more fundamental than an acquaintance with this or
that particular subject.

Before we can hope to agree upon the kind of education that
will best fulfil the functions ascribed to it we must have a clear
idea of the objective. What do we want our educational system
to achieve ? If we are satisfied with things as they are and wish
to preserve them so, the less change we introduce the better.
If we are satisfied that every man should learn "in whatsoever
state he is therewith to be content," we need no new kind of
thought. But if we wish our education to provide something
more than competent hacks, well versed in the three R's to meet
the routine requirements of commerce and industry, we shall
have to be prepared for changes in education, changes more
fundamental than a mere raising of the school-leaving age.
Remember, these people are to be trusted with the franchise.
However we limit their responsibilities in other respects, we shall
trust them to exercise whatever ability they have in the choice
of their rulers. The future of the country depends on its policy,
the policy upon the men in power, and the men in power upon
the choice of the children we are now educating.

According to Dean Inge : "The spirit of man does not live
only on tradition," and changes will come whether we will or not.
New and vigorous ideals must constantly infuse individuals and
communities and these will infiltrate into the educational system
and provide the basis of the training of the next generation. In
education we must always be looking ahead. If we are to
prepare the children of to-day for the world of to-morrow we
must know either what that world is going to be like or what
we should like it to be. Our worlds are mostly what we make
them, so that it is more profitable to know what we want the
future world to be than to hazard a guess as to what it is going
to be. The training that we give our children now will determine the shape of the world to come.

Are we ashamed of and disgusted by our slums ? Our
children must learn to abhor slums and whence the slums have
arisen. Are we horrified by the spectacle of an army of unem-
ployed, of whole towns dependent on Public Assistance, of areas
scheduled deliberately as "Distressed" ? Our children must learn
to be horrified too, but more, to question the system under

which these things can be. Are we concerned at national famine in the midst of international plenty ? Our children must learn that this is wrong, but more than that they must learn to ask why it should be so. Are we distressed by international lawlessness and the spectacle of two world-shattering wars within a quarter of a century ? Our children must learn to hate lawlessness and to believe that wars are not inevitable, but more than that, they must learn to enquire into the causes of law-breaking and war. Are we disturbed by the lack of interest now displayed in civic and political life ? Our children must learn to be honoured by the trust reposed in them, but more, they must learn to question the extent of the trust and how best they may fulfil it. And so we could go on, lengthening the list of what we may feel are undesirable aspects of our modern life and adding more of our ideals for the future until we have a programme of learning for our unfortunate children quite overwhelming, unless we can find some common denominator that will enable us to cover variety in simplicity. There is, fortunately, a common feature to all the problems that will confront our future citizens. They must learn to question, what questions to ask and, having asked them, how to provide an answer. The child who has learned this need fear no individual subject, since he will have evolved a basic procedure for handling all subjects. He will become objective in his outlook, restrained in his judgments, unbiassed in his opinions and unmoved by specious argument. In short, he will have learned to think.

But let us be aware that we know what we are doing. To teach children to think and to think honestly and fearlessly may be dangerous. It becomes a habit that is hard to break and the presence in our midst of any reasonable proportion of thinking men may produce a revolution ; not necessarily a bloody revolution, but a nonetheless effective one for that. There will be bitter clashes of opinion between the young idea thinking clearly and a generation who never learned to find for themselves, by the application of their reason, the answer to a single question. To encourage the young desire to think and experiment may be inviting trouble. The new thinking generation would hardly be likely to join the ranks of the many who are now prepared to take their opinions on the most important subjects ready-made. They will wish to be independent in their views and they may want independent leaders.

We ought not to be unduly worried by this form of revolution. The lack of independence of thought lies at the root of the easy propagandising of nations to their harm. When people cannot or will not think for themselves, others are in a position to think for them and to impose their wills upon them. When these others are unscrupulous or dishonest men the havoc wrought can be appalling, but it may be no less even when the operators of the propaganda machines are sincere if they are the possessors only of forceful personalities and not of the ability to think honestly and unswervingly. Sincerity in the leaders, though necessary, is not enough if they are incapable of perceiving the fallacies in their arguments and the inconsistencies in their policies. But when whole nations are half educated and can be easily misled by plausible arguments it is absurd to blame the leaders. They merely suffer from the same disabilities as their peoples.

It is no doubt already clear that we are about to suggest that more attention should be given to scientific instruction, or perhaps we should say, to instruction in science in the schools. Our children must grow up to be mature citizens. To do this they must learn to think, patiently and logically, with a set determination to reach a conclusion, and we submit that the best, in fact the only proved, pathway to this is through science. Proponents of the present educational system may perhaps point to examples of successful men and claim that the system as it stands is no bar to advancement. It might equally well be claimed that many who have received little or no education have also established positions and reputations for themselves. Personal success is so frequently a matter of personal character that it can scarcely be regarded as a criterion of the value of an educational system. Moreover, in achieving personal success many of these men have spread havoc around them by their ill-considered policies. The present state of world affairs can hardly be accepted as a justification of present education. We need more than an education that will not be a bar to personal advancement. We need a training that will be a positive assistance in any walk of life and the way to this is through the inculcation of a method of thought that has been tested, that can be applied to all subjects, to all people and to all circumstances. There is one such method and that is the method of science.

It is legitimate to ask at this stage why, if science possesses in fact the extraordinary attributes suggested, it has not long ago become the most important feature of education. Instead there are many still who deprecate the amount of science even now taught and who would resist any extension of its claims upon the already crowded time-table. There may be good reason for this antagonism, but we would do well to consider to what extent it is due to the deficiencies of science as a suitable medium for instruction or to the deficiencies of education in its employment. One thing is certain, whether it provides all that is claimed for it or not we cannot afford to neglect it. In a world surcharged with technical development such as ours, a knowledge of the sciences is a necessity, and this alone is a sufficient reason for including them among the subjects considered essential to the proper education of the nation's children.

It is just this possibility that it may be useful that has proved one of the biggest stumbling blocks in the way of better organised scientific instruction in the schools. Educationalists have ever been afraid of anything beyond the elements that might be considered useful knowledge unless it is directed to vocational training. There is even a degree of contempt for useful knowledge dating from Plato, whose teaching was associated with an aristocratic disdain of manual labour. Moreover, the idea that it may be useful has been so exploited that the kind of science taught has been deliberately directed to this end and has resulted in a mechanical instruction designed to be the preliminary to a career in technology. There are districts in the industrial north where science teaching does appear to have come into its own. In one village the local secondary school has never had a headmaster who was not a graduate in science, and the whole neighbourhood is knowledgeable of scientific matters. It would be pleasant to record that the inhabitants of the district and the products of the school stand out above their contemporaries for sagacity and sound judgment and that they have risen with ease to positions of public eminence. Alas, research has not proved this to be the case. They appear to be no better, if no worse, than their contemporaries in other localities. The reason is not far to seek and is common to other similar districts where scientific instruction has been accorded a high place. The instruction given has been dictated by local economic needs. There are chemical industries in the neighbourhood and it is confidently expected that a high proportion of the children leaving school will be absorbed automatically by the local industry for which they have been fitted by a form of instruction designed to prepare them to become technologists. The kind of science taught has been circumscribed by its very usefulness.

In this respect these districts mirror the general level of the teaching of the sciences. There is a standardised style for the

D

teaching of useful knowledge. Instead of the principles and methods of scientific thought, a mechanical routine has been devised. Chemistry and physics in particular are considered suitable for instruction, attention only recently being given to other sciences, and the teaching of these suitable subjects could hardly be better designed to kill any love for science that the struggling student might have had. The kind of teaching, for example, that occupies itself with facts concerning the elements and their compounds to the exclusion of the wider appeal of chemistry as a branch of scientific thought must often fail to grip or hold the youthful imagination. The scientific instruction that is necessary is not the suspect subject of practical utility now masquerading under this title, but a broad, humanitarian fundamental training in which the facts are illustrations chiefly of the principles and methods of thought employed. They will in all probability be scientific facts, not because historical or artistic facts are less important, but because they usually point the pathway to the truth more clearly, and it is the pathway that is important. Once the trail has been blazed it may be used over and over again to establish any sort of fact, and the scientific trail, although possibly not the only one, provides the only path that may safely be trodden by all. In addition, the facts of science are often simple and quantitative without being self-evident and so provide a practical aid to teaching by assuring to the student a definite conclusion to the argument and to the teacher a yardstick by which to estimate progress.

It cannot be stated too often nor emphasised too strongly that we are not seeking to produce technical experts. We have no ambition for our children to grow up to be physicists, chemists, zoologists or any other "ists," but mature men and women ; nor do we want them to grow up to be professional good citizens. Right-thinking men and women will automatically become good citizens though not necessarily docile ones. A training in science will teach the student to depend upon his own judgment in the assurance that, if the facts at his disposal are complete, his arguments are indestructible and his conclusions incontrovertible. He will learn too what is almost more important, that if his evidence is either incomplete or unsubstantiated, no amount of dogmatism will enable him to maintain his conclusions.

It is his mental equipment that will determine the child's future—not necessarily his success in the accepted sense of the word, but the quality of his life, its fulness, its sincerity. The facts he learns in school are of secondary importance and most of them will be forgotten by the time he is thirty. By that time he will have to look up the main products of the chief continental countries, the name of the highest peak in the Andes, the reasons for the Thirty Years War, Ohm's Law and the atomic weight of chlorine ; but if he has learned to think on correct lines, he will still be doing that automatically. Whether his problem concerns the weekly disposal of his wages, the disposal of the city rates or the spending of the national revenue ; whether it is the rule of his own home, his own life only, or the control in industry of the lives and activities of others ; whether he become a schoolmaster, parson, doctor or dock labourer, he will be forced by his early training to adopt the right attitude to his problems. He will be incapable of hazy thinking or woolly ideas. His mind, alert and capable, will analyse, probe and dissect every problem put to it. He will not be content with makeshifts or palliatives, but will seek to reach the root of his problems and, in the light of his reason, to act upon his deliberations with the sincere knowledge that they have been honestly wrought. This should be the aim of an education that seeks to produce, in the words of Thomas Sprat, "a race of young men provided against the next age."

It is not enough, however, to be convinced of the kind of education needed. Its success must depend to a large extent upon the progress that it makes in handling its own problems. The place of science in education and the science of education, although profoundly different, are thus closely linked. In any educational system that seeks to accomplish what we have outlined "the eyes should be trained to see, the ears to hear with quick sure discrimination. The sense of beauty should be awakened. The hands should be trained to skilful use. The will should be kindled by an ideal and a discipline enjoining self-control."[1] Sir Michael Sadler goes on to say that for the modern student : "His education should further demand from him some study of nature and should set him in the way of realising both the amount and the quality of evidence which a valid induction requires. . . . It should also, by the enforcement of accuracy and steady work, teach him by what toil and patience men have to make their way along the road to truth."

Are we then to have no teaching but science ? By no means. The problem briefly is, first, to provide a race of men not gorged with undigested facts, but so educated that all knowledge is within their grasp, all problems within their capacity to pass

[1] Sir Michael Sadler. Mather Lecture. *Liberal Education and Modern Business.*

judgment upon, and so trained that they can express themselves lucidly and appreciate beauty when they meet it. Scientific thought must be the keystone of education with this object, not so much because the modern world is technical, though this is important, but because it is only in the modern world that so powerful a mental tool has become available for the use of mankind. But science alone will not suffice. It is not enough for men to be able to reason and reason rightly, for life is not made up of a succession of rational *dénouements*. Men are creatures of the heart as well as the head and there are ideals to be lived up to and great causes to be fought for and superhuman sacrifices to be made, perhaps even for reason and sound judgment ; and men do not live and fight and suffer for abstract reasons. The results of the philosopher's lonely hours must be transformed into burning words, creative literature and moving music. So the student must not only be trained to pass judgment but to express clearly, forcibly and imaginatively his conclusions, so that others who have not delved so deep may catch the rhythm and be swung along in the vanguard of right thought. In one sense it is unfortunate that the sciences should need a set of defined terms for their own use. They are necessary because truth is nothing if not precise, but they do not sway men's minds. The student, then, must be taught to express himself lucidly, but more, he must learn to do so with grace. In this, unfortunately, few scientists possess any facility. And, since the world is a lovely place, he should learn to see the beauty around him and to appreciate that portrayed for him by the great artists whose eyes have glimpsed more than he can hope to see.

Such a programme needs the co-operation of educationalists. The scientist cannot carry it out, nor can the artist, nor the poet, nor the classicist and humanist in whose hands education has rested for too long. In recent years we have seen the beginnings of a race of professional educationalists. It used to be sufficient to have had some experience to be considered an educationalist. There have been great educationalists in the past, albeit they were taught empirically and by long and bitter experience, and their accumulated knowledge provides a basis for what, though it is early days, we may call the new science of education.

We are concerned here chiefly with the position of the natural sciences in education and it is no place to discuss the impact of the growing science of education on other subjects, but it is likely that much of what is true for the sciences will be equally true for other subjects. High schools and colleges must enlarge their scientific programmes, not to train scientists but to train men and women and to disclose to them new worlds of thought, and if science is taught in a manner calculated to stir the mind and train the reason it will need teachers who are capable of so presenting it and methods of presentation that have been tested and found not ·seriously wanting. For the latter we must have educational research, and the research must be carried out on scientific, which is the same thing as saying dependable, lines. Educationalists who can carry out the research must be trained. In fact, in education now we need what is always needed in any science, namely, men who have the divine gift of imagination coupled with the enquiring mind that forces them into research and men trained to appreciate, to apply and to teach the fruits of the researchers' labours.

Out of the seventeen universities in Great Britain, none offers a course in education leading to a first degree, though London, Durham, Leeds, Manchester and the Scottish universities offer higher degrees after two years of study and research in education. Where the entrants for these higher degrees are to get their preliminary training, either as educationalists or as researchers, seems a matter of little importance. All these universities train teachers, but education is more than a Teachers' Training Department run as a side line by the university for the Government revenue it provides. Every university should contain a school or faculty of education. Yet what is the actual situation ? Taking science as an example, although the picture is true in some degree at least for other subjects, we find that in the scholastic world a premium is placed on the possession of an Honours degree. In theory this appears to be an excellent provision, for it is obviously intended to encourage the best men with the highest academic honours to become teachers. In practice it works differently. In the first place, graduates do not decide after they have completed their university courses that they have a vocation for teaching. Instead, young men and women of seventeen or eighteen years are financed through their university years on condition that they agree to become teachers after completing the course. For many this is the only way of pursuing a university career, and, either in ignorance or with the avowed object of attaining a degree, many thus bind themselves who have no skill as teachers.

Nevertheless by this means they succeed in attaining their ambitions and securing for themselves the opportunity of higher education. The next step is that the young undergraduate enters upon his courses without reference to his future vocation, either on his part, his instructors' part or the part of the university. For four years, university lecturers and professors, picked men for the purpose, strive to impart a training in and a love for their own subject—chemistry or physics or biology—and in innumerable cases they succeed. It is their aim to create chemists, physicists or biologists, but not teachers. At the end of this time the now graduated scientist enters the Teachers' Training Department and spends a more or less intensive year in the study and practice of teaching, after which, if he is fortunate, he obtains a post as a teacher in a secondary school. What is to become of a man who, infused with the idea of chemistry or some other subject as a career, finds himself condemned never to proceed beyond the elementary stages with which he himself finished long ago ? It has taken four years to produce the first result and it is hoped to replace it by a new allegiance in less than one. It is a forlorn hope from the outset. Dissatisfaction must arise and that spells disrepute for the teaching profession. If this means anything, it is that at all costs the student must put his teaching first and his special subject second. He should leave the university fired with enthusiasm as a teacher and with a firm enough belief in his calling to carry him through its inevitable disappointments.

But how can all this be put into practice ? Simply and effectively. Students who are destined for a teaching career should enter the faculty of Education and should receive a degree in Education. The present two-year course could readily be expanded to meet the needs of the utility teacher and a special four years course in which his specialist subject could be taken to pass degree standard could be designed to suit the specialist teacher and to lead to an Honours degree in Education. The Honours degree in a special subject is utterly wasted so far as the needs of a secondary school teacher are concerned. The nobility and desirability of his chosen profession are the elements to be cultivated in the budding teacher and this cannot be accomplished in one year nor built upon another foundation. This problem is not a new one to which thoughtful men have given no consideration. Lord Eustace Percy wrote in 1930 : "The problem is . . . no longer how to ensure that the intending school teacher shall be caught up into the main stream of university life and education, but how to dig for him at the university a special channel, broader if somewhat shallower than the all too narrow canals which lead to an honours degree. It is becoming evident that if the secondary schools and the more advanced central schools are to be staffed in the future from the universities by ear-marked physicists, chemists and historians, while other senior schools continue to be staffed from the training colleges by ear-marked teachers, it is the latter type of school rather than the former that may come to represent the traditions of a liberal education."[1]

We do not need and we should not want specialists to the extent of an Honours degree in our schools. What is needed is men and women who have appreciated the methods of science and have been trained to pass them on ; teachers who are enthusiastic about teaching and whose reward is found in seeing the avid grasp of principles and methods rightly taught, in place of the dreary and unwilling acceptance of facts, figures and unassimilable details now forced upon children by a backward society in the name of education.

With men properly trained in education we could then begin to explore their possibilities as original investigators or practising teachers. As in other subjects, a selection may be made which will permit of a certain number pursuing a training in methods of research to be followed by employment either in the educational faculties of the universities or in some other field where scope for research is afforded. If there is to be more science in our schools it must be taught, not so much as a factual survey, but rather as an intensely human theme with the methods and procedures of the scientist in the elucidation of his problems always in the forefront. And for this we need men trained both to impart their knowledge and to investigate and expand the borders of the science of education.

SCIENCE AND RELIGION

"Though I have the gift of prophecy, and understand all mysteries, and all knowledge, and though I have all faith, so that I could remove mountains, and have not charity, I am nothing."

ST. PAUL.

SCIENCE and religion share with each other the distinction of being the two fundamental intellectual activities of mankind. All

---

[1] *Education at the Cross Roads.*

other aspects of intellectual activity seem to be compounded of these two ingredients in various proportions. They may be valuable or even indispensable to the peculiar circumstances of this civilisation, but they are not fundamental. Religion and science alone appear to possess this characteristic ; religion which co-ordinates, studies and produces a philosophy of human behaviour, hopes and fears, and science which synthesises a philosophy of the surrounding world—the world of force and matter. They both pose questions, the answers to which are not easily apparent, and both set out to seek the answers to riddles that draw closer and closer together as they are pushed further and further back, and unless there is some unknown boundary that lies between them they must come nearer and nearer to each other as each step is taken into the unravelling of causal forces. They start from opposite poles and there is a boundary now set by science which will not proceed beyond the physically demonstrable so that the realm of the spirit is barred. Accepting nothing on trust, science takes an infinite number of details and builds them up into a composite and, where possible, simple picture of the underlying and directing forces. Religion, on the other hand, armed with an already prepared explanation of primal causes, proceeds to apply it to the details of life and living. It is inevitable that the two should sometimes overlap. Medicine, employing the physical sciences to the full, is often materially assisted by outright religion—faith being the substance of things hoped for.

It is equally inevitable that they should sometimes clash, and the relation between them has been a stormy one with a history of dogmatic and tempestuous displays on both sides. Feelings have run high in the past and there is even now an undercurrent of passion between them. True, some theologians profess to be convinced of the right of science, while many scientists, no doubt, are satisfied that there is something in religion. Unfortunately it is difficult for them to find a common ground. The more arrogant claims of science have prevented religiously inclined men from considering it calmly and dispassionately. They have hastened to the defence of their religion and, securely barricaded behind its infallibilities, they have condemned science out of hand. Scientists, on the other hand, find it difficult to accept any sort of infallibility or finality and so have been contemptuous of the apparently flimsy barricade.

In the clash of thought religion is at a disadvantage. First because it is more difficult to be convincing about claims which may appear somewhat nebulous compared with the physical evidence of science and, second, because it rests upon an infallible revelation which is at once a source of weakness as well as its strength. Its strength because it provides an unassailable fortress to which to retire and a source of weakness because it permits no changing of ground. Science is not so hampered by stability of its theories. When clashes come it is a war between mobile guerillas and an entrenched army. It is, however, an army that cannot in the nature of things be dislodged and any victory to the scientific forces can only come as a result of sallies of the army from its defensive positions.

If, however, the arrogance of science has prevented religiously minded men from paying heed to it, its confidence has provided an excuse for others to evade what their consciences dictated to be a duty. A criticism here and a denial there and the whole fabric of faith has been destroyed for those whose consciences and inclinations pull in opposite directions. This is probably most true for those who have been bred in a religious atmosphere with its aura of prohibitions. With an inbred dislike of religious observances, the refuge of science is a safe one against the assaults of conscience. There are undoubtedly men whose outlook forbids them to indulge in religious extravagance, but there are others whose inclinations have been fed on the second-hand assertions of the sciences.

To what extent the arrogance of science is justified may be judged in the light of the questions put to Job by God.

> "Who measured out the earth ?—
> do you know that ?
> Who stretched the builder's line ?
> What were its pedestals placed upon ?
> Who laid the corner-stone,
> when the morning stars were singing,
> and all the angels chanted in their joy ?
> Who helped shut in the sea,
> when it burst from the womb of chaos,
> when I swathed it in mists,
> and swaddled it in clouds of darkness,
> when I fixed its boundaries,
> barred and bolted it,
> saying, 'Thus far and no further !
> Here your proud waves shall not pass' ?

> .   .   .   .   .

> "What path leads to the home of Light,
> and where does Darkness dwell ?

Can you conduct them to their fields,
and lead them home again ?
Have you found out the fountains of the sea ?
Have you set foot upon the depths of the ocean ?

. . . . .

"Have you grasped the earth in all its breadth ?
How large is it ?   Tell me if you know that.

. . . . .

"Can you bind up the Pleiades in a cluster
or loose the chains of Orion ?
Can you direct the signs of the Zodiac,
or guide the constellations of the Bear ?
Can you control the skies ?
Can you prescribe their sway over the earth ?
Can you send orders to the clouds,
for water in abundance to be yours ?
Can you send out the lightning on its mission ?
Does it say humbly to you 'Here am I' ?
Who taught the feathery clouds,
or trained the meteors ?
Who has the skill to mass the clouds
or tilt the pitchers of the sky, ——"[1]

These are, by and large, questions which the physical sciences
have set themselves, but to-day, some two thousand five hundred
years since they were asked, the answers to most of them are
still incomplete.

The nineteenth century was the most prolific in scientific
invention that the world has known, and having achieved its
greatest successes in the study of the physical and material
worlds it is not to be wondered at that its early incursions into
the fields of philosophy should have followed a mechanistic path.
It does not follow that a mechanical basis is the only one for a
successful philosophy, and science, which may have presented a
rival philosophy of life and one that is wholly acceptable to some
minds, has not necessarily superseded all other philosophies.
This it may do only if all other philosophies are incompatible
with scientific thought.

We must therefore, bearing in mind the controversy that has
long existed between science and religion, ask ourselves whether
the religious philosophy is incompatible with a scientific intelligence.   Herbert Spencer wrote : "Of all antagonism of belief

[1] Job, Chapter xxxviii. *A New Translation of the Bible.* James
Moffatt.

the oldest, the widest, the most profound and the most important is that between religion and science. It commenced when recognition of the commonest uniformities in surrounding things set a limit to all-pervading superstitions." The linkage of religion and superstition is significant. To what extent are religion and superstition synonymous ? The Jews had a religion handed down for generations and claimed as originally Divinely given to a peculiarly favoured race. Nevertheless there must have been a strange admixture of superstition with it by the time Christianity burst upon its scene. It was certainly commercialised, as witness the money-changing and the pigeon selling in the Temple courtyard for the sacrificial offerings. The Greek religion, with its gods and goddesses subject to very human frailties and forming a very similar hierarchy to human society, led St. Paul, who must have been familiar enough with superstition to be hardly moved by it, to exclaim to the Athenians : " I perceive that in all things you are too superstitious." It was in the presence of such religion that science was born. Every inch of the way was contested between the two antagonists. Christianity with its claim in turn to a Divine origin grew, from its humble start among fishermen, slaves and wandering disciples, into a highly organised, lucrative and powerful religion absorbing many of the pagan festivals and beliefs, with one Pope at Rome and another at Constantinople. And the Christian church came into violent opposition with science. Established religion had decided that the earth was the centre of the universe ; that all things revolved around it, including the sun. Galileo, however, believed differently and had sound observation of facts to support his view. The first round was won by religion and the world remained the centre of a whirling universe, but nobody now believes this to be true and it is difficult to see that there was ever any Christian reason for the belief or that Christianity has suffered a setback by the recognition that the earth is only one, and not perhaps the most important, of many planets circling around a central sun which itself may be only one of many suns.

The modern controversy that has raged between science and religion may not be so very far removed from that contest between Galileo and the Church. The opponents of science may well examine their beliefs to see whether they fight for a fundamental of faith or for a traditional dogma the loss of which would leave faith unchallenged. According to Joshua the sun stood still in the heavens through Divine intervention on behalf of the

Israelites, and the story has lost none of its meaning nor its savour because we now know that the sun does not, as it appears, traverse the skies from horizon to horizon.   The picture remains the same and indeed a modern artist in words might, for all his superior knowledge to Joshua, use the same imagery to describe the event.   The essence of belief in God is not, after all, a question of physical measurement and explanation, nor does it rest upon a dogmatised creed, but upon faith and personal experience. Sir Wilfred Grenfell, the Labrador doctor, wrote : "Christ called for faith in Himself.   He never called for intellectual comprehension.   He sent out to preach His gospel men who had not any creed or any intellectual faith, only a dumb sort of faith that Christ was more than man."

It is inevitable that intellectual comprehension ceases somewhere, but we do not discard immediately all that we do not fully understand.   If we are to do this we must cease to use electricity for heating, lighting, power, telegraphy, wireless and the thousand and one important and useful applications to which it has been put.   We must indeed give·up the use of all forms of energy, for although we may measure and describe, probe and dissect, analyse and peer ever further into the unknown we are still ignorant of the nature of force itself.   It is a phenomenon which we recognise and which we reduce to tameable proportions, but we do not understand it.   The chemical energy released or absorbed by reactions which we use to our daily benefit is an unknown quantity so far as understanding is concerned.   The catalysts which promote reaction and make life possible are beyond our understanding.   We must no longer accept the fruits of the earth and we must prevent the continuation of our kind and the breeding of animals—if we reject that which so far is outside our comprehension.   But we do not reject these things. We use them and seek by patient enquiry to understand them more.   We accept them as part of the pattern of experience.

So with religion we must take experience where we find it and, accepting it as a facet of life, seek to understand it.   That religious experience is real cannot be denied.   There have been too many saints and martyrs for it to be in doubt.   There are too many people even to-day whose lives are regulated and, to all outward appearances, irradiated by an inward spiritual experience. It is easy to criticise the Christian church.   It is harder to explain it.   Its professors are not perfect.   It is, in its present organisation, open to criticism.   But there must be life there to support through two thousand years and the direst persecution so stupendous a growth. There must be some foundation for faith. The philosophers, statesmen and even scientists who have been prepared to stake everything on the truth of the possibility of religious experience have not been easily misled. Experiences are not peculiar to a few and if we are honest we must enquire diligently into the causes of their faith. We must be candid, serious and unbiased in our enquiries, and still keeping before us the excesses that have been perpetrated in the controversy between science and religion we would do well to recall Herbert Spencer again, who wrote : "In the proportion as we love truth more and victory less, we shall become anxious to know what it is that leads our opponents to think as they do."

The resources of scientific thought can be brought to bear upon religious experiences as upon other phenomena and, as with other phenomena, they may explain and illuminate it though they cannot change its nature. Explained or not, natural phenomena go their own ways. The techniques of music and literature and art may be improved by careful scientific investigation, but their natures are unaltered and we may have perfection of technique and yet lose the essence of artistry. Forty years ago W. L. Watkinson realised this and wrote : "The great master is inspired, lifted up, carried away by a force which is not himself. The orator is himself a mouthpiece ; the painter is himself a pencil ; the musician is himself an organ-key responsive to the touch of an invisible finger. Materialistic writers attempt to take the mystery out of us, to reduce us to the simplicity of sewing-machines, and to explain categorically the genesis, the process, the ending, of all our doing ; but the masterpieces of art, literature and language declare that there is something more in the world than mud and motion ; that there is a supernatural element in man, the inspiration of a higher world."

At the same time that christians are examining the fundamentals of faith, scientists might well be evaluating concepts which have proved a stern battle-ground for religion, endeavouring to see to what extent the controversy has been inflamed by the absence of the scientific spirit. The scientist who dispenses with the spirit of humble enquiry is riding a steed with a hard mouth that is likely easily to get out of control and to throw his rider. The materialism which formed the basis of the attacks on religion is dying fast. Science, by her own investigations, is retracing her steps and does not now speak with the same arrogance as heretofore. The structure of science is constantly changing and the materialistic outlook of the nineteenth century is already superseded in the twentieth. It is no longer possible to be dogmatic on the blindly mechanical forces activating nature. Man himself no longer seems to be a purely fortuitous being, mechanically controlled by forces over which he has no power, and when the time arrives for a conclusive statement on the scientific belief of the century there will undoubtedly be provision for previously unsuspected spiritual harmonies.

Deep feeling is as real as deep thought and they must walk hand in hand if the greatest benefit to mankind is to result. If scientific thought is to be applied to the problems of the day its direction must be dictated by men of goodwill. Indifference to the results that science produces and the ruthless application of the principle of force must give way before an idealism which does not deny the weaknesses of mankind, but which also through faith believes in its ultimate destiny. And where is such faith best to be found but in the Christian religion, which has shown repeatedly what changes can be wrought in individual lives when the defeat of self and selfishness is accepted as the chief victory— a victory won at nobody else's expense. With it comes the conviction " that truth is always better than falsehood, that kindness is always better than cruelty, service than arrogance, love than hate." These are theorems as fundamental as the laws of science but less austere, warmer and more personal, assuredly the logic of the heart rather than of the head. Both are necessary to the man of affairs. It is not sufficient to be well-meaning ; nor is it sufficient to be scientifically correct. Goodwill must be informed by scientific thought and scientific thought must be illuminated by goodwill.

These two fundamentals of thought are curiously similar in many ways and in their similarity is a further proof, if one were needed, of their basic character and intimate relation to one another. They resemble one another in the effects they produce in their sincere practitioners. It has already been suggested that the true scientist is not likely to be an easy individual to fit into life as constituted to-day. Professor L. W. Grensted, in a little book recently published, has described the outlook of true christians and says of them : "Such men and women will not necessarily be very easy to fit into the ways of the State, either in its inner social structure or in its outward relation to other States. They will be restless in their own spirits, and very like a bad conscience in the community as a whole. They will be continually inconvenient to themselves and to everybody else, even while they are foremost in willing service, since they will be sensitive to every failure of justice and will see the lack of love in many things that ordinary folk take for granted. The beggar at the door, the underpaid and ill-housed farm labourer or artisan, the child in a slum and the growing boy without a reasonable prospect of work, will fill them with an acute discomfort for which the prosperity of other sections of the community, possibly including themselves, will afford no compensation at all. The State which has encouraged them in this freedom of spirit will be restless too. But in its restlessness it will also be progressive." The tender conscience of the true christian allied to the clear analytical thought of the scientist would revolutionise both public and private life. But perhaps we are not anxious to see this revolution. Too many accepted, if not acceptable, standards might be overthrown.

They resemble one another, too, in the distinction that we draw between precept and practice. In neither case are the professors invariably good advertisements for the profession. There is a grandeur of quality about the vitality of the Church that is not often reflected in the lives of so-called christians. The ethics of the Christian religion, viewed dispassionately, are beyond reproach, but how often are christian men and women shining examples of the philosophy they embrace? And all through these pages we have perforce drawn a careful distinction between science and the professional scientists. The truth is that the perfection of the counsel rises above the attainments of the exponents.

Religion and science are related, too, in that they are both opposed to the natural predilections of men. In this, indeed, lies the reason for their inconvenience to the world and the shortcomings of their professors. The natural inclinations of men, which run to self-seeking, sloth and ease, and which manifest themselves in deceit, dishonesty and ruthless acquisitiveness, are diametrically opposed to the discipline and sincerity of both Christianity and science.

The christian and the scientist should not, therefore, have far to go to meet each other and for the welfare of the world they must meet. The christian in the world needs the guidance of scientific thought to save him from descending into sentimentality, but the scientist in the world needs too the sweetening influence of the christian outlook to soften and render acceptable the harsh outlines of his incisive thought and to ensure that what we know to be good predominates over what we know to be evil. For we must adopt a standard and we may confine the scope of scientific thought to a low or even a wrong standard. We have said earlier that science does not distinguish good from evil and we believe this to be true. By its inexorable logic it shows the course of events and predicts the result of behaviour, but it does not attempt to say whether that result is right or wrong, desirable or undesirable. A decision must be taken on this when the end is foreseen. Here, in fact, is the crux of the whole application of scientific thought. Having explored the possibilities of a particular line of action leading inevitably to an undesirable result it still remains possible to prosecute it with scientific thoroughness and to ensure the efficiency of the end. Moreover, what may be a desirable end to an individual, such as personal advancement, may perhaps be achieved only at the expense of others. It has been the constant thesis of these pages that much evil is wrought by inability to foresee the logical outcome of behaviour but it is lamentably true also that even when the evil is predicted it may still be pursued.

A decision must be taken on the end to be sought. On what criteria is it to be based ? It was Adam Smith, I believe, who laid down a principle for distinguishing good from bad in the economic field which may have a general validity. His principle was based on a scientific outlook, for it presupposes the ability to deduce the consequences of action. He suggested that the morality or not of any course of action could be settled by a consideration of its result in the world if it were followed by everybody. If the result is patently bad then the action is evil ; if not, then it is, at worst, neutral and may be good. For example, it is bad for one man to kill another, for were all allowed to kill their neighbours the result would be annihilation. Defence by individual superiority of armament is bad since in the nature of things it is impossible of universal achievement. This principle does not in itself provide an absolute distinction between good and evil. Itself it must lay down as a condition that the total result must be good or bad, though we might define its terms further by the acceptance of some such criterion as "good is that which produces more happiness than pain." Who is to be the arbiter of good or bad, or even of happiness ? Who is to decide that annihilation is bad and on what grounds, or that universal achievement of defence by individual arms were it attainable is good ?

It seems that good and bad must be accepted as axiomatic descriptions, understood by all and questioned by none. It is not possible to evolve a logical criterion by reason alone ; an initial act of faith is always necessary ; an act of faith in a fundamental law of the universe. The scientific method is a necessity of life, but alone it is not enough. It must be practised by men of goodwill whose faith in the end they serve is deep-rooted and firm. The hope of this civilisation lies in the discovery of men in whom the accurate thought born of a scientific training is combined with the faith that moves mountains in all that is good and operates through a sense of service that outstrips all hope of reward.

SCIENCE AND LEADERSHIP

"To complain of the age we live in, to murmur at the present possessors of power, to lament the past, to conceive extravagant hopes of the future, are the common dispositions of the greatest part of mankind ; indeed the necessary effects of the ignorance and levity of the vulgar. Such complaints and humours have existed in all times ; yet as all times have not been alike, true political sagacity manifests itself, in distinguishing that complaint which only characterises the general infirmity of human nature, from those which are symptoms of the particular distemperature of our own air and season."

EDMUND BURKE.

THE relation of science to the fundamental activities of men and to their common experiences is important, but not more so than its impact upon their qualities, and of all the qualities that will be called for in the world that lies ahead leadership comes first. Whatever kind of future we may fashion for ourselves will depend upon the leaders we elect to follow or allow to be imposed upon us. We cannot yet see what the new world will be like, but sufficient is emerging through the fogs of the immediate present to make us aware that it will be full of problems the solutions of which will have no precedent. Many of the problems are directly traceable to the effects of science, and this may mean that their solutions will have to be devised with its help. In a world increasingly dependent on the achievements of science for its material well-being the responsibility of scientists for the direction of affairs becomes unique, and there are many who hope that, by accepting the implications of their responsibility, men of science will be able so to direct policy and to mould opinion that circumstances may be controlled to serve men's best desires

E

rather than that their highest hopes be circumscribed by uncontrollable circumstance.

Are there any grounds for believing that such a hope can be effectively realised ? Hopes and fears alike are often reared upon unsubstantial foundations composed for the most part of the desires or frustrations of the hopeful or fearful. Both alike are to be eschewed by wise men unless some solid ground can be established upon which to pin them, when they may become the springboards of action. In one sense scientists are already the leaders of the new age. It is the technological advances resulting from their skill that permit the definition of policies on a wide stage, although this often waits upon something other than the provided means. The means to an end does not alone ensure the adoption of that end. When the Lead Chamber Process was first introduced for the manufacture of sulphuric acid there resulted a widespread nuisance from the escape of obnoxious nitrous fumes into the air. In 1827 Gay-Lussac showed that this disability could be overcome by absorbing the offending vapours in strong sulphuric acid, but this improvement did not come into general use until 1860, just one year after Glover had demonstrated that the trapped gases could be returned to the chambers, to bring about a considerable economy in the use of the comparatively expensive nitrate. Until then the means that lay to hand was of no advantage. But influencing progress by such indirect means is not the positive direction that moulds circumstances to the needs of man.

To do this requires qualities in and opportunities for scientists that will make them leaders of thought and activity. It has been possible so far to dissociate science and scientists, but now that the qualities of men are being called in question it is no longer possible to divorce the philosophy from its practitioners. It is now necessary to examine what kind of men professional scientists are and whether they can measure up to the standards to be set for them. If the hope is to be realised that in difficult fields where others have failed scientists may succeed, certain qualities in the men themselves are essential. In particular they must possess an objective outlook and an administrative or executive ability. The first of these has long been accepted as characteristic of scientists, but the second has not usually been associated with scientific attainment.

Frankness compels the admission that even the first of these qualities, so commonly considered the prerogative of men of science, has not invariably been the outstanding characteristic it is supposed to be. The scientists we know are, after all, not automata, but men, and as such they still possess and exercise in greater or less degree the foibles of men. It is not always the case that they carry the unbiased, unprejudiced outlook of their science into everyday life. Like other men they sometimes become partisans whose judgment is conditioned by their adherences, and, looking around in the restricted scientific world, there are too many evidences of bitter controversies to allow one to accept comfortably the theory of scientific objectivity. When they disagree among themselves they do so with all and more of the fervour of other men. Even so they are a curious race. Accustomed to work and wait patiently for recognition they seem sometimes to possess the worst characteristics of spoilt children. Occasionally they exhibit tempers and jealousies unsurpassed in any profession. Perhaps it is desirable to continue to distinguish between science and scientists, for science surely could wean men from the petty exhibition of human passions in which scientists sometimes indulge. Yet to deny them humanity would be to kill for ever the possibility of their becoming leaders. Let it be said, then, that while they cannot claim perfection the probability that scientists will be objective in their judgments is greater than with most men, and that, in fact, the frequency of its occurrence has been sufficiently remarked to establish a tradition of unprejudiced outlook.

The possession of administrative ability has not, however, been recognized as a scientific attainment, but rather the opposite. The tradition here, if there is one, favours the absent-minded, impractical or doctrinaire outlook. There is a general feeling that scientists are unsuited to the control of enterprise though they may provide the tools for others to do so. Leadership has naturally been for the most part denied to them. It has been supposed that the humanities provide the suitable training for leaders and few scientists have challenged the tradition by taking up public service in a manner that would show them to be leaders. By reason of this a vicious circle has been completed. Scientists have not sought office and so have confirmed the judgment of public opinion in denying it to them.

Nevertheless there ought to be men of science who possess administrative abilities combined with technical skill. They may not be Fellows of the Royal Society nor Doctors of Science—criteria by which they may be judged by their fellows—but men who in addition to their scientific qualifications possess that further quality which, for want of a better word, we term leadership. This ability to inspire confidence and to obtain desired ends cannot be restricted to one class of men. It is unreasonable to suppose that a classicist or a lawyer or an historian will perforce possess it while a physicist or a chemist will not ; yet the former have expected and received administrative responsibility while the latter have not.

There are reasons for this, of course, which must be sought chiefly in the training that scientists receive. Is it not possible that there is something in a scientific training that tends to unfit a man for executive positions ? May it not be accompanied by serious drawbacks ? We had better face the fact that scientists are different from other men—different from the men they are to lead if they become leaders. They think differently and they speak a language of their own. More than most men they find their greatest interest in their profession and they are obsessed with quantitative considerations which are foreign to the majority.

The scientific mind is a specialist mind. It is trained to think according to a certain pattern and that pattern is different from the normal. That it is the most successful pattern we know does not help. It is its difference, the fact that it is set apart, that it appears aloof, which prevents it from being either generally acceptable or generally understood. Its very logic gives it an austerity that is displeasing to the ordinary man. " Because" and "therefore" are the guiding principles of the scientist. He claims to perform no miracles and accepts few mysteries. Difficulties he admits but none that cannot be surmounted by time and patience. Where other men circumvent he has an obstinate tendency to persevere in his efforts to overcome. Scientists are like the "sandhogs" of America who, out of sight, strain every nerve to blast and pump their way through any obstacles to build a tunnel that may provide an easy passage for others. But they are out of sight. Not one in a million that uses the Hudson Tunnel to get quickly, safely and comfortably from New York to New Jersey gives a thought to the sandhogs whose labour made it. Scientists have also an accuracy of thought that is alien to the woolly cerebration that sometimes passes as humanistic, but it is not the accuracy of arithmetic. That belongs to the bookkeeper, but the accuracy of the scientist springs from a clarity of thought that is directive and anticipatory, so much so that they may appear to have been born before their time. It is not that they may be wrong, nor even that they are sometimes misunderstood, but that they are often premature that keeps the scientists in a class apart.

The peculiarity of the scientist's mental outlook is reflected in the singular precision of his speech. St. John Gogarty, in his *Tumbling in the Hay*, describes a one-time tutor at Trinity College, Dublin, thus : "McNought has got a life sentence from Science or, rather, to it. He is surrounded by frozen, dead and inelastic words that have only one meaning. Stone walls do not a prison make but scientific terms do," and he goes on to say that "a word should be like a Chinese ideogram admitting of hundreds of interpretations, permitting full play to the imagination, adumbrating Truth in its every shade and meaning." No scientist would subscribe to this. Words to the scientist are valueless until they have been defined beyond all misunderstanding, until in the light of the above extract all the vigour and liveliness have been defined out of them and they have been reduced to mere skeletons of their former selves. In the search for precision, scientists have invented whole new vocabularies for themselves, introducing new words and giving exotic meanings to familiar ones. The familiarity of ordinary words which endears them to normal men makes them unpalatable to the scientist. They have connotations and associations which are foreign to the exactitude that he wishes to convey. Moreover, even in the use of the dead-alive exactitudes that they permit themselves, most scientists are unskilled. At its best, scientific jargon is hardly beautiful or stirring, and in the hands of men of science it may become almost barbarous in its uncouthness.

In one sense scientists are too modest. They do not take the time nor the trouble to explain in a simple, lucid manner the conclusions they have reached. There are, of course, even among scientists, men like Jeans and Eddington, but what are these few among the many—the many sometimes who, not possessing the gift of exposition, are tempted to despise it. There is too great a tendency among scientists to be content with the virtues of the progress of knowledge and to let its dissemination look after itself. But the dissemination of knowledge is almost, if not quite, as important as its growth. The power of forceful expression is the stock-in-trade of every successful public man, be he politician, lawyer, schoolmaster or parson, and it is a power that can be acquired. Every student of the arts obtains a facility with words that, without a natural aptitude, is denied to students of the natural sciences. It is not simply sufficient to have driven the tunnel of knowledge a few feet through the solid rock of ignorance, nor to have ideas of how to drive it further and faster. The knowledge must be bruited abroad, the ideas flung into the broad stream of human consciousness ; and without a facility with language this cannot be done. Lesser ideas and antiquated knowledge if better presented will surpass deep thought and hard-won new knowledge.

The beauty of language can stand alone as can the dignity of truth, but they can be matched to make a rare combination. It is unnecessary to discuss whether it is preferable to have ideas without words in which to express them adequately or a form of expression without ideas upon which to drape it ; whether a body is more important than its clothes : but it may be suggested that the danger of seductive language is the same as that of seductive clothing. Just as the deficiencies of an unattractive figure may be hidden and an altogether false impression of beauty be given by carefully designed clothing, so fallacious arguments and loose thinking may be so clothed in acceptable language that they take on an unexpected verisimilitude. What science badly needs is a band of interpreters who will link its truths with litera-ture. To disseminate the knowledge of science we need novels and plays in which science takes the place of the happy ending, and poetry which sings its beauties.

Until they have learned to express themselves, scientists will continue to be wallflowers at the world's quickstep. Unfortu-nately, all the signs point to an increasing gulf between scientists and the world in which they live, for science in its passion for exactitude is becoming increasingly mathematical and seemingly divorced from reality. What would once have been hailed as a beautiful piece of work is fast becoming a sound piece of mathe-matics, and the truths of science are taking on the appearance of a passionless procession from one formula to the next. There is no longer even a concern for the physical significance of the formula. It has become in itself the successful end of the search. Not in this way are the qualities of the scientists likely to be recognised. Scientists, in their writings, are too truthful and that makes them dull. In their modesty, they have no use for hyperbole. The idea of exaggeration for effect is foreign to them, but the resemblance to Greek thought pointed out in the first chapter can be carried too far. Scientists are unsatisfactory also in the difficulty they find in giving a direct and categorical answer even to the most simple question. Their yeas are so seldom yea and their nays so seldom nay. They for ever hedge them about with conditions and only within the given conditions will they pledge their word. It is in this, to many of them seemingly unimportant, ability for exposition that scientists are most lacking and something will have to be done about it if they are to be accepted as leaders. They must learn to write and to speak so that people listen to them if they are to be understood and appreciated. Splendid isolation gets neither nations nor men anywhere.

The isolationist attitude of the scientist is not dictated, however, solely by modesty nor by snobbery nor even by lack of ability. He has a peculiar faculty for immersion in his career which occupies his thoughts to the exclusion of other interests. He represents the first fruits of William Morris's ideal that every man should enjoy his work for its own sake. He has, as a result, no end beyond the work itself, no desires beyond its satisfactory performance and no satisfaction apart from it. He is concerned with the performance of the task in hand to the exclusion of the objective to which it is directed. It is for this reason to some extent that the products of scientific research are not exploited by scientists but by men without scientific training. As a class, scientists have recently been claiming more responsibility for the application of their work, but how can they achieve this when, as a class also, they are scarcely interested in the uses of their science provided they are left in peace to enjoy their limited laboratory pursuits.

This detachment from ends and concentration upon means is a further peculiarity distinguishing scientists from ordinary men. Where the ordinary man works for personal advancement, and finds in his success sufficient reward for unloved drudgery, the scientist works for the love of science (and is usually expected to) without thought of its possible repercussions upon himself. This is a truly distinctive mark of the genuine scientist and is the basis of scientific, though perhaps not worldly, success. As a career, science from the very first demands the whole energy of its followers. As students they spend long hours in the laboratories in addition to the quite abnormal amount of reading required to master the unfolded knowledge of the past and to keep abreast of the constantly increasing discoveries of the present. For a scientific career all else must be sacrificed. The young enthusiast entering upon it takes upon himself the marriage vow : "Forsaking all others."

There is, too, a secondary danger in this absorption in their work that sometimes shows itself when by force of circumstances a scientist is "elevated" to a position of eminence. It is that, having immersed himself in a world of his own making, his judgments in a larger world of others' making may be conditioned by his limited experience. He should be the last man to rely entirely upon experience, since he has learned to think, but the temptation to argue by analogy with experiences with which he is familiar is one that does not assail the scientist only and he may be forgiven if in such unfamiliar fields he is tempted to lose faith in his simple precision of thought. With experiences often so far removed from immediate practicality as are some specialist scientific fields, susceptibility to this temptation is fraught with more peril than usual and, since he is already suspect, one faulty judgment is more condemned in the scientist than half a hundred in other men. The disservice to science as a whole is correspondingly great. One well-known scientist, a Fellow of the Royal Society, to whom an important executive position had been accorded, admitted privately that he preferred the laboratory to the office because in the former he could predict the direction of events. Naturally he could ; he had spent thirty years of his life in the laboratory, but to rely upon that experience to determine the behaviour of men was inviting disillusion.

Yet, after making allowances for all the disadvantages so far enumerated under which the scientist labours there seems to be a smaller incidence of administrative ability among them—compared with lawyers, historians or classicists—than one would like to see. There seems to be more than a possibility that just as science has emasculated their language and restricted their experience, so it may have confined their appreciations within impregnable barriers. Training in science, the physical sciences in particular, with its insistence upon quantitative factors, may perhaps deaden the mind to considerations of quality, the sense of which marks out the administrator. The very aspect of his training that teaches the scientist to think exactly and precisely may rob him of the administrative ability to use his gifts to the best advantage. In the realms of human relationships, finance, economics or politics there are no blacks and whites such as the scientist has become familiar with, but many varying shades of grey. They cannot be assessed in the quantitative terms to which he is accustomed, but must be nicely balanced without exact knowledge.

And if by its nature science limits the appreciation of quality, how much is this intensified by the recent habit of youthful over-specialisation. The early narrowing of the fields of endeavour to a single furrow which is steadily ploughed deeper is fatal to the budding leader. It may be a good thing as a specialist to know more and more about less and less, but it is killing to the administrator who must know at least enough about more and more.

On the face of it, the chances of a scientist becoming a successful administrator look remote. Circumstances and even his own training and experience seem to militate against the possibility. Yet there have been scientists who have achieved recognition in other fields of endeavour and it is possible to trace their ultimate success to their attachment to the principles and discipline of their training, for all the disadvantages of science are just ripples on the surface of a steadily flowing tide. They are but superficial and may be avoided without losing the benefits of scientific thought which of itself can bring success.

Scientists have been successful business men, as witness Lord Cadman and Lord Kelvin among others ; while in Dr. J. B. Conant, the successor to the mantles of Eliot and Lowell in the Presidency of Harvard University, and in Sir James Irvine, the Principal and Vice-Chancellor of the University of St. Andrews, there is evidence that the scientist need not fall behind his fellows in the world of educational administration. Indeed, it is not too much to say that the modern provincial universities in England were built by scientists and achieved their success by the trust they reposed in science. The University Colleges of the nineteenth century owed their growth not to a renaissance of accepted learning but to a liberal recognition of the natural sciences. Their principals, men of vision and wisdom who established them well, were, for the most part, scientists chosen in the first place as suitable leaders of science colleges, but they were hardly, I imagine, expected to build them up into universities of the standing of Birmingham, Bristol or Liverpool.

At least one modern ecclesiastic—the Bishop of Birmingham—is a scientist ; the Commander-in-Chief of the Canadian Active Service Force—Lieut.-General McNaughton—graduated with a degree in science from McGill University and is in times of peace the President of the National Research Council of Canada, and many other modern soldiers, sailors and airmen have high academic qualifications ; the composer Borodin was professor of chemistry at St. Petersburg, while the General Musical Director of the National Broadcasting Company of America came from the University of Pennsylvania with a degree in chemistry. The British Broadcasting Corporation owes its structure largely to one man—Lord Reith, who as a young engineer took on the task of guiding the footsteps of the new departure in public enterprise. Whatever criticism may be levelled at Lord Reith, and some has been, let it be always remembered that he was one of the new ruling class of technicians who, without precedent to guide or experience to rely upon, built the B.B.C. from nothing to a vast corporation and trod successfully a wary path in the full blaze of public opinion for nearly twenty years.

The blazing light of public opinion has poured upon other men of science too, who have emerged from the dim fastnesses of scientific remoteness and successfully withstood its glare. It is pleasing to record that, though they may not have been numerous, there has never been, so far as the writer is aware, any breath of scandal attaching to a scientist in public life. This, in spite of the fact that Lavoisier, the father of modern chemistry, was guillotined while Commissary to the French Treasury. Perhaps the greatest of them was Benjamin Franklin, printer, journalist, scientist, diplomat and virtual founder of the United States of America. He was one of the five men who drew up the Declaration of Independence in 1776 and one of three Commissioners who in the same year were sent by the new republic to solicit the aid of France and the sympathy of Europe in the struggle for independence. According to Dr. John Bigelow,[1] it was largely to the reputation which had preceded him as a man of science that the success of his mission was due. He lived on to be a member of the convention that framed the Constitution of the United States, which after more than one hundred and fifty years still persists as the basis of life in the new world.

But Lavoisier and Franklin belong to history. What evidence is there that scientists can take their place in the modern world of affairs ?

Not everybody, perhaps, considering the meteoric rise to fame of the Minister of Aircraft Production, is aware that Sir Stafford Cripps is a scientist. It is as a lawyer that we think of him, but before embarking upon a legal career he had won a Science Scholarship from Winchester to Oxford, worked with Sir William Ramsay in his laboratory and managed (during the last war) the Queen's Ferry explosives factory.

In an earlier chapter it was suggested that scientists would not fit easily or comfortably into the modern world. These men, who have proved themselves, have been no exception to that rule. They have without exception followed the dictates of their own conscience and reason without regard to the consequences and what consequences followed when Franklin the

---

[1] *The Complete Works of Benjamin Franklin.*

British subject threw in his lot with the rebellious colonists. With the background so briefly sketched above it is not surprising that Sir Stafford Cripps should be expelled from the Labour Party after stating as a profound conviction : "I am loyal to my principles, not to any Party machine." In Benjamin Franklin, one scientist ushered in a new world. There are stirring changes ahead. Is it possible that another scientist, thinking profoundly and untrammelled by specious loyalties, may yet prove a leader capable of orientating them to the future benefit of mankind ?

Let us not doubt that scientists can be successful in other walks. Then what has marked out these men from their contemporaries that they have achieved what we may legitimately hope might be attained by many more if science possesses the value that has been attached to it in these pages ? In no case have they sought honour or office. In this they have been typical of their class. Authority and power have come to them without their seeking. We need not expect that every scientist will be a leader. Something more than science is needed for that. The men who have won renown have not won it as specialists. They have pursued more than one course, have set their specialist knowledge on one side, but used the rigorous procedure that it has inculcated in the often trackless wastes which lay just beyond their carefully tilled fields. But with these examples before us and in the light of the value of a scientific training we have a right to expect, nay, to demand, that every leader shall be a scientist. Leaders there must be and scientists they should be, for if the dead hand of their specialised training can be lifted they have peculiar gifts to bring to the task.

How are the handicaps imposed by science simultaneously with its benefits to be surmounted ? It is the task of education to provide the leaders of the future whether they are scientists in whom the quality of leadership has been unimpaired by their training or humanists who have been trained in the scientific method. In the meantime the nation must seek out those who have, unaided, overcome the obstacles placed in their path by their profession. They will, as a rule, be found among men who have not confined themselves to a narrow field of specialisation, men who have been something besides scientists, and this gives a clue to the kind of training that education must provide for the scientific leaders of the future. It is not a knowledge of the structure of matter, the laws of terrestrial magnetism, the foundations of atomic theory, not an acquaintance with biology, astronomy, chemistry or physics that has brought famous scientists renown as public servants, educationists or business men, but a first-hand acquaintance with the principles of science in whatever specialised field they may have been applied. They have been men with other interests than science which have broadened and matured their outlook on a life subject to the same principles that permeated their profession.

In general terms it does not matter whether we train scientists in such a way that their potentialities as leaders are not lessened or whether we superimpose a scientific outlook upon laymen. From whichever point of view we look at it, it seems evident that there is scope for an educational training symbolised by that provided in some American and Canadian universities under the title of Liberal Arts and Science. By this means we may hope to produce men for public service who combine the advantages of both groups, men who are exact in their thinking and cultural in their outlook. It is by some such method that education must solve the problem created by the presence in the community of scientists who cannot express themselves, who are, in fact, nothing more than technologists. It is likely to be easier to superimpose a scientific training upon laymen than to inculcate into scientists the additional knowledge of affairs that they must have. But although education in principles is not so protracted a task as the production of a technically competent scientist and may be taken as the easy road, there is a danger that a superimposed training may become a superficial one without the practice that makes the scientific method second nature.

We are more concerned here to provide a class of scientist who can take his place without fear among the administrators and arbiters of policy than to impose scientific thought upon unwilling humanists. It cannot be denied that this presents great difficulties. The demands of his profession, both during his training and in its subsequent practice, cut the scientist off from many of the normal pursuits which of themselves strengthen the man of affairs. Long hours in the laboratory are not conducive to the study of men and this concentration must be counteracted. It is not now uncommon for a schoolboy to spend the later years at school specialising in scientific pursuits. For the potential leaders there must be no specialisation before entering the universities, and the entry should not be at too early an age. The advantages of a broad basis of knowledge carried to an advanced stage cannot be over-estimated, and the practice of concentration on a limited field of study immediately after matriculation is a dangerous development of modern education. Specialisation at school is probably bad in any subject ; in science it is suicidal. But naturally students specialise to-day. The university courses in scientific subjects are too short. They have been stabilised at three or, in some cases, four years, for a considerable period during which science has been continuously expanding. There is, therefore, no time for the absorption of the philosophy of science since so much is occupied in digesting the plethora of knowledge. Students of science need more time in which to develop and mature their appreciative and critical faculties.

One American innovation has been referred to already. More time spent at the university would allow of the adoption of another American custom that has much to recommend it, particularly for students of science, namely, the introduction of one or more subjects not strictly necessary for the attainment of degree standard and preferably unconnected with the degree course. Each student could select for himself from a list of acceptable subjects those which appealed most to him and without the necessity of pursuing these supplementary studies to degree standard. The object of the additional studies would be to provide, in training at least, that more complete outlook that makes for the appreciation of quality. They would help to prevent the complete absorption in one line of activity that is now the greatest danger for the youthful scientist.

More time would also allow students of science to take a greater part in communal activities than the demands of the present congested syllabus permit. In pleading for an extended course, the basic assumption is made that thereby the science syllabus would take on more the aspect of a cultural education than a vocational training. The majority of present students of science are drawn from the ranks of those to whom it is imperative to earn a living, and an interest in chemistry, physics, botany or some other scientific pursuit coupled with the short course, the scholarship or grant-in-aid facilities and the reasonable certainty of some kind of post-university job, attracts them into the laboratories where they may be ground to pattern between the twin millstones of a voracious study and economic pressure. How fallacious is the idea that the present short course will be sufficient to fit the student for his profession may be judged from the increasing frequency of higher degrees. It is becoming increasingly difficult to obtain posts in industry without the additional experience denoted by a Master's or Doctor's degree.

To lengthen the courses would be a recognition of an already existing situation and would give an opportunity of improving considerably the training that the student receives. There are few who do not agree that advanced degrees are obtained with a minimum of benefit to the students and there has been some advocacy of abolishing the Ph.D. altogether. It would probably be preferable to improve its character by the addition, to the present requirement of an elementary piece of research, of advanced courses either on specialised topics or on the more speculative aspects of science.

Every opportunity should be given, too, for students of science to travel and rub shoulders with all kinds of people. In theory this idea is subscribed to universally, and fairly comprehensive arrangements make it possible for outstanding students to spend a year or two abroad, but in practice this usually resolves itself into the student fitting himself into another very similar routine to that to which he has already accustomed himself and from which he returns to his own country with the dignity of foreign travel but lacking its advantages. And what incentive is there for him to do otherwise ? What sort of career would be open to a modern young scientist who decided to adopt the line of action taken by Descartes and spend the years from age twenty-three or so to about thirty roaming over Europe ? Yet there is much to commend this for potential leaders as Descartes' reasoning indicates. In his own words : "As soon as my age permitted me to quit my preceptors I entirely gave up the study of letters and, resolving to seek no other science than that which I could find in myself or else in the great book of the world, I employed the remainder of my youth in travel, in seeing courts and camps, in frequenting people of diverse humours and conditions—and above all in endeavouring to draw profitable reflection from what I saw. For it seemed to me that I should meet with more truth in the reasonings which each man makes in his own affairs, and which, if wrong, would be speedily punished by failure, than in those reasonings which the philosopher makes in his study." The modern student who took this as his model would find that his contemporaries had passed the novitiate stage into which he would be expected to enter and that he had unfitted himself as the minor cog that he must be content to be in the industrial machine. If he had also courage he might find that the gates to leadership were not so fast barred against him as against his seemingly better situated colleagues who had not given way to such fanciful notions.

In the meantime, such steps must be taken as are within the bounds of practicality. It is not possible to provide opportunities for foreign travel for all the promising young scientists, but something might be accomplished by forbidding or, at least, discouraging a man from taking an advanced degree in the same university from which he graduates. Let him go elsewhere and meet new men with new ideas and so let him realise that much of what his professors have told him is not so much a gospel as a point of view. No student worth his salt in any university worth the name has any more to learn from his professors by the time he has graduated. This does not imply that he has absorbed all the facts that they can teach him, but these, after all, he can get from books and other men just as easily and he has reached a stage where a point of view is more important than a host of facts. For four years, perhaps, he has consistently absorbed one viewpoint and it is high time that he realised that there are others ; so let him proceed to another university.

Furthermore, not only should the student be encouraged to take himself elsewhere to contact new viewpoints, but every opportunity should be taken to bring new viewpoints to him while still in his most formative years. The practice of making appointments to the staffs of universities from within cannot be too strongly condemned. The life blood of scientific faculties must be constantly renewed from outside. More interchange and still more interchange is necessary so that the student world may appreciate that the scientific world is more than the stagnant pool it often seems in any single academic spot.

What will thus be done for the student of science ? Will he automatically become an administrator ? Or has he been taught to be an executive ? No ; but he has been given opportunities to surmount the obstacles that the study of science might place in his path. His future remains his own. Only the germ of leadership within him can make him a leader. It cannot be implanted by his teachers or his training, though either may kill it, but in their wisdom his teachers can, if they wish, provide a training that will, without destroying his scientific thought, so widen his horizons that he may go out into the world and gather for himself the experience without which there never was a successful administrator.

But when all is said and done, is it worth it ? Do we want scientific men in control of affairs ? Abundantly yes. They are the answer to the growing criticism of scientific advances summed up by Cicely Hamilton in *Life Errant* in her description of the "malignant power of the chemist, the aviator, and the engineer," and ultimately in her conclusion that "ignorance of the powers and forces of Nature is a condition of human existence." So long as the power to exploit the advances of science for base ends remains in the hands of men undisciplined by its rigours and ignorant of the forces with which they carelessly toy, so long will scientific knowledge remain "a power of destruction, illimitable destruction placed in the hands of a fallible emotional humanity." There must be scientists among the leaders of the new world, for they alone understand the machines they create and can control the powers they unleash. But it is less the specialist knowledge of the scientist that is in demand than their specialist thought, and this must be made increasingly available to all. The stone age, the bronze age and the iron age have come and gone. Each had its own peculiar needs. The spread of the enquiring mind, the healthy scepticism and the basic honesty of scientific purpose is the first requirement of the new scientific age.

THE END